If You Tame Me

Understanding Our Connection with Animals

IN THE SERIES

Animals, Culture, and Society

edited by Clinton R. Sanders and Arnold Arluke

If You Tame Me

Understanding Our Connection with Animals

Leslie Irvine

FOREWORD BY MARC BEKOFF

TEMPLE UNIVERSITY PRESS PHILADELPHIA

Temple University Press, Philadelphia 19122

Published 2004
Printed in the United States of America

♾ The paper used in this publication
meets the requirements of the American National Standard
for Information Sciences—Permanence of Paper
for Printed Library Materials, ANSI Z39.48-1984

Library of Congress Cataloging-in-Publication Data

Irvine, Leslie.
If you tame me : understanding our connection with animals /
Leslie Irvine ; foreword by Marc Bekoff.
p. cm. — (Animals, culture, and society)
Includes bibliographical references (p.) and index.
ISBN 1-59213-240-5 (cloth : alk. paper) —
ISBN 1-59213-241-3 (pbk. : alk. paper)
1. Dogs—Behavior. 2. Cats—Behavior. 3. Dogs—Psychology.
4. Cats—Psychology. 5. Human–animal relationships.
I. Title. II. Series.

SF433.I78 2004
636.7'0887–dc22 2003070299

2 4 6 8 9 7 5 3 1

Contents

Foreword: To Know Them Is to Be Them

Most human beings form close attachments with at least some nonhuman animal beings (a.k.a. "animals"), usually companion animals (a.k.a. "pets") such as dogs and cats. Often those who are not sure about how they feel about other animals have a sort of love–hate relationship with them, even if they cannot articulate why. That is how closely we are tied into the lives of other animals, whether we like it or not. I have been studying various aspects of the social behavior and cognitive and emotional capacities of animals for more than three decades, and I begin my studies by asking the deceptively simple question, "What is it like to be a _____?" where the blank can be filled in with one's animal of choice. When I enter into their worlds—for it is essential to try to understand and appreciate their worlds and their worldviews—I become the other. Thus, in my studies of coyotes, I am a coyote; likewise with my research on dogs and birds. Our identities become blurred, and the borders that many construct become permeable. Leslie Irvine writes mostly about companion animals, and in my long-term research on dogs, I see that these remarkable beings not only follow their hearts but also their noses, ears, and eyes. They can be victims of their senses and of their unrelenting curiosity and boundless love, and that is why we love them, why we identify with them, why they are us.

I love Leslie's book. It is accessible and at the same time well researched and scholarly, filled with "hard science" (what I call "science sense") and anecdotes (one of the two nasty "A" words and what some pejoratively call "soft science"). Anecdotes are basic to most, if not all, of the sciences, for it is usually stories that motivate further empirical or experimental research. Moreover, it is important to remember that the plural of anecdote is data.

The second "A" word that raises eyebrows and hackles is "anthropomorphism." As Leslie and others have pointed out repeatedly, there are no viable alternatives to being anthropomorphic, and if used carefully and biocentrically (taking into account the world of the animals) anthropomorphism can motivate further empirical work. We are anthropocentric because, as humans, we have to be to make sense of the behavior of other animals. Critics of anthropomorphism offer timeworn and, frankly, boring claims that being anthropomorphic makes the lives of other animals more, or too, human. I think they are wrong, for by being carefully and biocentrically anthropomorphic, we can make the lives of other animals more accessible and at the same time take heed of what we are doing.

Leslie has titled her study of the close relationships we have with animals after an episode in one of the most enchanting children's stories of all time. However, it is really one of the most enchanting stories of all time. In *The Little Prince,* the adult must learn that sometimes it is best to accept life's mysteries. In addition, it is an animal, a fox, who reveals that what really matters in life can be seen only with the heart, not merely with the eyes. Antoine de Saint-Exupéry dedicates *The Little Prince* to his best friend, but he is careful to clarify that he means to dedicate it to the child that his now grown-up friend once was. The same sense of honoring and recapturing the wonder and possibility of childhood guides Leslie's book. Children are curious naturalists, and they readily accept that animals are thinking, feeling beings. Children understand that animals can sense and share our emotional states. As we grow up and become "educated," we learn that this is nonsense—that animals do not really feel emotions; that they are robots or automata. Now we know that this portrait of animals is not only demeaning but also patently false, for it flies in the face of scientific data that show that many animals have rich and deep emotional lives. And, as Leslie points out, their patterns of social communication can be very complex: A growl is not a growl is not a growl.

Fortunately, some of us refuse to be trained, refuse to let supposed objective and value-free science get in the way of our really learning about the animals with whom we share Earth. Leslie is squarely situated in the group that at once respects scientific data but also knows that there is more to the study of animals than pure science. Although this book draws on research Leslie conducted during the past five years, the underlying premise that other animals are emotional and feeling beings has driven her since childhood. As she relates the story of her encounter with a baby elephant, readers will see the first indication of her curiosity about animals' interior lives.

However, this is not a children's book. It brings together solid evidence and theory in ways that will convince the staunchest "grown-ups" to abandon their notion of animals as machines (if indeed they still stubbornly hang on to them). Leslie's research involved careful observation of behavior, on the part of human and non-human beings. It is a rigorous study. Yet it blends rigor with compassion, social responsibility, and heart into a recipe that the Austrian scientist Anton Moser calls "deep science." Over the years, some of my colleagues have called me "flaky" because I talked to my late dog, Jethro, and watched and listened for the ways he responded to me. Let there be no mistake that his form of communication—his language, dare I say—was deeper and richer than most words I know. Perhaps some of Leslie's colleagues in sociology will have similar thoughts, thinking that this book falls far outside the realm of "legitimate" sociological concerns. Some might say that, given the many human problems in the world, studying animals is a waste of precious time. However, no one is outside of nature. How we relate to other animals influences how we view ourselves and relate to other people. Moreover, just as we have learned that there is a correlation between cruelty to animals and cruelty to humans, it is important to remember that compassion for animals can make this a more compassionate world overall. Caring might spill over into sharing.

The main point of Leslie's book is that people's relationships with companion animals are what they are because animals, like people, have selves. My research on selfhood shows that animals other than primates have evolved selves that work for them in the social milieus in which they spend most of their time. By looking only among those who share our ability to use language, perhaps we have been looking for self in all the wrong places. Cognitive ethology has also shown us that many diverse animals have rich mental and emotional lives, much of which remains a mystery. However, Leslie argues that one of the clues to the mystery appears in the research on the interior lives of infants. Like animals, human infants cannot use language, and yet few would deny that they display the basic elements of what we think of as a sense of self. They recognize mother and know that she is not them; they share smiles, laughter, surprise, and other emotions; and they can initiate their own movements in pursuit of a goal, such as reaching for a bottle or a toy. Although human infants go on to acquire language and consequently develop the kinds of selves that are useful for humans, Leslie argues that the pre-linguistic elements of selfhood exist among animals, too. She uses vivid examples to illustrate how this selfhood becomes apparent to their human companions in the course of everyday interaction.

But this book does more than offer a theory of animal selfhood. It calls for action: proactive compassionate activism, a practice that can heal the wounds we inflict on other animals and the wounds we suffer when we do so. Our big and old brains make us the most powerful beings on Earth, and we can do just about whatever we choose to other animals. Thus, we bear incredible responsibilities to be moral beings and to step lightly with grace, humility, respect, compassion, kindness, generosity, and love. We can be and do no less if we are to live in reciprocal harmony with other animals. We really are that powerful and ubiquitous. Perhaps activism can be viewed as a type of payback: We are paying back other animals for all that they selflessly give us, whether we know it or not. The late Martin Luther King, Jr., once claimed that apathy is equivalent to betrayal. I have argued that indifference can be deadly for other animals whose lives and voices depend on our goodwill. If you find the evidence in this book convincing, it will lead you to consider how you think about and treat animals. In her research at an animal shelter, Leslie found that many people who supposedly love animals also treat them as commodities. People "trade animals in" when they fail to behave in ideal (that is, in human) terms. People give up on animals when they grow up, grow old, grow too large, or become otherwise inconvenient. More to the point, many people do not make the effort to honor animals' sense of self, and they consequently deprive themselves of a rich and deeply satisfying relationship. As Leslie lays out the ethical positions that follow from her argument about animal selfhood, she clearly and concisely establishes the differences between animal welfare and animal rights. Many people think these are the same, but they are not. Leslie's claim that animals have rights raises many difficult issues but bravely gives voice to those who cannot speak for themselves.

The possibilities that this book opens are endless. They are also challenging and frustrating. How we view other animals informs how we view who we are in this huge and diverse world. We are unique, but so are other animals. And when we carefully parse the criteria that have been frequently used to separate "us" from "them"—tool use, language, art, culture, feelings, consciousness—we find ourselves on thin ice, for none shows that we represent some sort of evolutionary discontinuity. Charles Darwin argued for evolutionary continuity, that physical and mental differences among animals were differences in degree rather than in kind. We ought to take seriously his challenging views. Leslie, too, suggests new ways of understanding, appreciating, valuing, and loving the other beings with whom we share this planet. Venture into this text with an open mind. More important, read it with an open heart. For as the fox told the little prince, the heart will tell you what is important. When you listen to your heart, it will draw you one step closer to creating a world in which animals are appreciated and loved for who they are: individuals with their own personalities, pain and suffering, joy, and love to offer.

Mark Bekoff

Acknowledgments

I am indebted to Marc Bekoff for conversations about animal behavior and for reading an earlier draft of the manuscript. Thanks are also due to Clint Sanders for reading earlier versions and providing invaluable encouragement. Steven and Janet Alger's guidance improved the manuscript significantly. I am grateful to the Algers, as well as to Jackson Galaxy, for important insights into human–cat interaction. The friendship of Brian and Melanie Pelc provided support and laughter along the way. The staff at The Shelter never seemed to tire of my questions. Janet Francendese of Temple University Press guided the manuscript through every stage of production. Through it all, Marc Krulewitch believed in the project, and more important, he believed in me. For that, I am truly grateful.

Portions of this book are adapted from previously published articles: "The Power of Play," *Anthrozoös* 14 (2001): 151–60, with permission, International Society for Anthrozoology (ISAZ); and "A Model of Animal Selfhood: Expanding Interactionist Possibilities," *Symbolic Interaction* 27 (2004, forthcoming), with permission, University of California Press and the Society for the Study of Symbolic Interaction.

If You Tame Me

Understanding Our Connection with Animals

Introduction: The Fox's Wisdom

This is a study of how the animals who share our lives influence who we are. It is based on several sources of data collected during three years of research. Most of this research took place in my work as a volunteer at a humane society that I refer to as "The Shelter." I also interviewed people who were adopting and surrendering animals, and I observed them as they came to look at homeless dogs and cats. In addition, I observed and interviewed people at community dog parks and drew on my own reflections about a lifetime of living with animals.

The title of this book comes from Antoine de Saint-Exupéry's famous story *The Little Prince,* long a favorite of mine. In chapter 21, the prince, after arriving on Earth, crosses deserts and climbs mountains but finds no friends to stave off his great loneliness. To make matters worse, he finds a rose garden, and the sight of the vast numbers of roses makes him realize that his beloved yet troublesome rose, who is back on his planet, is not unique, as she had adamantly insisted she was. He misses her deeply and begins to cry. Then the prince meets a fox, and he invites his new acquaintance to play. The fox explains that he cannot play because he is not tamed. The prince asks what "tamed" means, and the fox says that it means, "to create ties." This does not help the prince, who says aloud, "To create ties?" The fox goes on to offer an explanation that translates well to the experiences we have with companion animals:

> "That's right," the fox said. "For me you're only a little boy just like a hundred thousand other little boys. And I have no need of you. And you have no need of me. For you, I'm only a fox like a hundred thousand other foxes. But if you tame me, we'll need each other. You'll be the only boy in the world for me. I'll be the only fox in the world for you."[1]

The fox's wisdom speaks to the questions I examine in this book. This is a study of how the animals with whom so many of us share our lives become the only ones in the world for us. I set out to explore how we develop a sense of self in relation to animals, and I found that, to participate in the process of self-creation, animals have selves, too. Granted, animal selves differ in degree from the selves we possess. Animals do not worry about what they will make of their lives; nor do they write autobiographies. Nevertheless, the selves of animals enable them to participate in relationships with us, and relationships in turn maintain, strengthen, and sustain selves. Throughout this book, I examine how our interaction with animals makes various aspects of animal selfhood available to us.

Skeptics will scoff that this is merely anthropomorphism, but I disagree. Although I treat the topic of anthropomorphism later in this book, I will tip my hand here to say that if people simply projected onto animals the qualities they wanted them to have, then any animal would make a good companion. I did the research for this book in an animal shelter, and anthropomorphic projection would make the adoption process much like ordering a pizza. You would not even have to meet the animal. You could simply fill out a form and ask for, say, a gray, female cat, and take the one you get. Even if you were able to meet the animals, making a match would simply be a matter of selecting a cat or dog whose appearance you liked. I show in later chapters how the cat or dog who looks right is very often all wrong for a particular person. In observing people meeting and adopting new animals, I learned that they sought one with whom they felt a "connection," to use the word I heard people use themselves. Appearance and behavior mattered, to be sure, but not as much as the connection. The term suggests that there must be something with which to connect. And, indeed, I argue that this is the animal's sense of self. Animals have elements of a core self that becomes present to us through interaction with them. Other researchers have found evidence of this core self in infants; thus, it does not require the use of language. Moreover, the elements of the core self correspond with what William James and George Herbert Mead described as the "I," or the subjective sense that has been so difficult to study. Finding the "right" animal is a matter of finding one whose core self meshes with ours. We recognize the elements of the core self in the animal, and the process of doing so confirms the self within us.

When interaction develops into a relationship, additional dimensions of animal selfhood become available as the animal's intersubjective capacities become apparent. For example, relationships present opportunities for humans and animals to share intentions and feelings. We share them not only in an "I know what you know (or feel)" way, but also in a more complex "*I* know that *you* know that *I* know *what you know*" manner. Over time, this cannot help but shape our identities. In other words, animals participate in the creation of our identities through many of the same processes that other humans do. They challenge our interactional abilities. They share many of our emotions and our ideas. They can surprise us, and yet they act predictably. In addition, they contribute to our histories, so much so that one woman I interviewed suggested that, instead of translating dogs' lives into human years, we should measure our own lives by the animals who have populated them.

If you are still skeptical about this, let me offer another reason not to be. The agenda I had in mind when I began this project was to study people who were adopting animals for the first time in their adult lives, if not for the first time ever. I thought that studying new relationships with animals as they unfolded would give me a window into animals' roles in the construction of selves. The animal would constitute something new and different in their lives and thus would form a point of comparison of "before" and "after" selves.

In theory, the plan sounds good. In reality, however, I was able to locate only two people who fit this category. Everyone else had had dogs or cats since childhood or as children.[2] To be sure, there were animal-free gaps here and there, such as the years in rental apartments or while married to an allergic spouse. Overall, though, people who had animals had more or less always had them. Moreover, a majority of people who were adopting animals already had other animals in their household. My inability to find a "virgin" population of first-time guardians suggested that, once animals entered people's lives, they stayed there. The continual presence of animals suggests something about their role in selfhood. Because people who had animals as companions had done so for much of their lives, animals must provide something that becomes indispensable. Once accustomed to having that "indispensable something," doing without it seems unthinkable.

I will admit up front that this has been my experience. I have always loved animals and can count only about three years when I have lived without them. My current household contains four cats and two dogs. I began thinking about the questions that orient this book when I was about eight years old. A petting zoo had come to a local shopping mall. In addition to the usual goats and sheep, the zoo had a baby elephant. I do not want to imagine the horrors that forced this creature to a mall in western New York, but my encounter with him forever changed me. He was about a foot or two taller than I was, so I had only to tilt my head slightly to look in his eye. He was chained by one foot to something heavy. The sight of the chain troubled me immensely. He had been rocking back and forth in the way bored, frustrated elephants do until I approached him. I remember seeing his ribs expand and contract in a sigh. I touched his hide and felt his trunk and the bristly hairs on his chin. I ran my hand over his ears. But what I remember as much as what he felt like is what I felt in his presence. He stopped rocking and leaned toward me just slightly. We were both very quiet. I could hear his breathing. I realized that here was another being who preferred contact to being alone, who had a history, albeit a mostly dreadful one, and who had feelings that could change from restlessness to something like contentment. My father let me spend an inordinate amount of time with the elephant, and I remember having thoughts along the lines of, "This is not a stuffed toy in the flesh. This is not Dumbo. This is another being, like me, and yet not like me." Coming home from the mall, I could not wait to see our dog. I took him to my room and, in a silent conversation, told him how important he was to me and thanked him for who he was. As the fox had put it, he was the only dog in the world for me.[3]

At that moment, I had begun "minding animals," to use Marc Bekoff's apt term. As he explains, the phrase has two meanings. It means "minding" them, as in caring for them. It also means attributing minds to them, or, as he puts it, "wondering what and how they are feeling and why" (Bekoff 2002, 11). I am interested in both meanings of the phrase, and I can trace my awareness of this interest back to the time in the mall with the elephant. The opportunity to study this would not come for many years. When, in the early 1990s, I began to explore topics for a Ph.D. dissertation, I considered but abandoned the idea of studying relationships with animals because I could find scant sociological literature on which to build. However, I am pleased that, over the decade, that has changed. The growing list of sociological studies includes Clinton Sanders's work on living and working with dogs (Sanders 1990, 1991, 1993, 1994a, 1994b, 1999, 2000; see also Robins et al. 1991); his work with Arnold Arluke on the ways we think about animals (Sanders and Arluke 1993; Arluke and Sanders 1996); Arluke's work on the use of animals in research (Arluke 1991, 1994); Clifton Flynn's studies of animals and family violence (Flynn 1999, 2000a, 2000b); Corwin Kruse's research into gender and animal-rights activism (Kruse 1999); Jennifer Lerner and Linda Kalof's study of animals in advertising (Lerner and Kalof 1999); David Nibert's work on animal rights (Nibert 1994, 2002); and Steven and Janet Alger's analyses of the culture of cats (Alger and Alger 1997, 1999, 2003). In 2002, the journal *Society & Animals* celebrated ten years of publishing interdisciplinary work in human–animal studies. The same year, the American Sociological Association recognized a section devoted to Animals and Society, which meant that several hundred sociologists had indicated their interested in the topic by joining the section in formation. To be sure, there are detractors who scoff, "Animals and society—good grief! What's next?" Nevertheless, a critical mass of sociologists has decided that the social world does not consist only of humans.

This book is a work of interpretive sociology aimed at building empirically grounded theory that can expand the notion of what constitutes "the social." However, I intend the arguments contained in these chapters to be useful beyond the field of sociology, and I hope that the readers will include people who are not sociologists. My goal in this book is to provide a theory of animal selfhood. Most of us who live with and love animals know that they have feelings, preferences, personalities, and other, similar characteristics. Sanders and the Algers have studied people's relationships with animals and documented that attributing selfhood to animals is indeed commonplace. However, it is also commonplace to dismiss these attributions as something silly and sentimental in which "animal lovers" indulge but that have no verifiable basis. The task I set for myself was to provide such a basis, or to learn what capacities in animals allow us to attribute selfhood to them. Here is an overview of where the discussion will go along the way.

The first three chapters discuss how we—by which I mean human beings—arrived at a historical moment in which we could begin thinking seriously about animal selfhood. The broad brush I use to portray vast expanses of time will make historians wince, but my intention is to offer a reasonable summary of the literature. In Chapter 1, I trace the questions of how and why we came to have close relationships with particular species of animals. The chapter opens with a history of dogs' and cats' domestication. Although much of what we know is open to debate, I focus on the most convincing current evidence of how dogs and cats came to be the companion species of choice in human societies. I then examine several prevalent explanations of why we continue to want them in our lives. Among these are the notion that animals are surrogates for human relationships (the deficiency argument); that animal companionship accompanies wealth and leisure (the affluence argument); that relationships with animals allow people to demonstrate power (the dominance argument); and that evolution has designed human beings to want to be close to animals (the biophilia argument). I maintain that any one-factor explanation is bound to fail, largely because our relationships with animals have meant so many different things over time. Chapters 2 and 3 then examine some of these meanings. In particular, I trace the concepts of "animal," "pet," and "companion animal," concentrating on the social and cultural factors that helped bring each term into currency at certain times. I argue that our relationships with animals, on the cultural level, have as much to say about human beings as about animals themselves. In other words, working out who is an "animal" and what rights and privileges come—or, more accurately, do not come—with that designation has much to do with defining who is fully "human." "Pet," in turn, reveals the working out of class relations.

Chapter 4 shifts the focus of the book from the theoretical and historical to the empirical by drawing on research in the adoption areas in the place I refer to as The Shelter. (Readers who are curious about my methods will find details in the appendix.) The discussion begins with the observation that most people who come to The Shelter do so only to look at the animals, not to adopt. The chapter goes on to analyze the appeal of looking at homeless animals, focusing on two dimensions that point in the direction of animal selfhood. The first dimension refers to the "trying on" of possible selves, similar to window shopping but different in that animals, as agentic beings, involve structural changes in people's lives. A second dimension refers to the aesthetic experience of looking at animals. I draw on aesthetic theory and social psychology to argue that looking at animals establishes them as coherent, physical beings. This, in turn, hints at animals' subjective capacity, which consequently confirms our own sense of self.

Chapter 5 examines first interactions between adopters and potential adoption candidates. The chapter establishes the principal "types" of adopters: those who come in search of animals of a certain breed, size, or color, and those who simply want the "right" companion dog or cat. Ultimately, the attraction hinges on whether the person feels an emotional connection with an animal. The analysis shows that many of the social-psychological theories and concepts of liking and attraction come into play during initial interactions with companion animals.

It is easy to dismiss this feeling as anthropomorphic projection, but that does the experience a disservice. For if the sense of connection with an animal came solely through anthropomorphizing, then people could project almost anything they want onto an animal. Adopting an animal would involve finding a cat or a dog of the right color who could obey the right commands. However, there is usually much more to making a good match, and the right cat or dog is sometimes all wrong for a particular person. What prospective adopters seek is a sense of what the dog or cat is like. In other words, they are trying to get a preview of the animal's self. In specifying how animals communicate this to potential adopters, the chapter lays the groundwork for the subsequent discussion of animal selves.

Chapter 6 sketches out a model of the self that does not depend on language, which animals can therefore share. There are many different ways to conceptualize the self. These include the soul or spirit of religion, the "inner child" of pop psychology, and the more academic portrayal of the collection of roles we perform for an audience. There is also a growing focus on the self as narrative, a model that I have used in previous work (Irvine 1999, 2000).[4] In short, no one can agree on exactly what the self is. Postmodern scholars even argue that the concept itself has become irrelevant (see Gergen 1991). In this view, technology's compression of time and space has so multiplied the possibilities for interaction that discussions of a single, "true" self have become outmoded. However, everyday experience tells us that the lack of theoretical or conceptual agreement about what the self is—or whether it is—means little; there is a "there" there. We have a very real experience that we are or we have a sense of self that has a central position in daily life.

In examining the structure of interaction between people and animals in the adoption areas, I found that different types of interactions shared the theme of the self. In other words, the ways that people interacted with animals suggested that people saw animals as contributing to a sense of who they are. Although many things contribute to who we are—art, music, hobbies, nature—animals do so in a different way. Our relationships with animals are more like relationships with other people than with objects. We see animals as having minds, emotions, preferences, and other things that indicate subjectivity. However, the question of what enables us to see them as such raises important questions for social psychology, which relies on a language-based model to explain subjectivity among people. The reliance on language eliminates a considerable amount of interaction as a source of information that contributes to selfhood. If factors beyond spoken language matter, and I argue that they do, then animals can participate in the creation of human selfhood. For animals to do so, they must themselves be subjective others. But how can we sense the subjective experience of animals when they cannot tell us how they feel and what they think? I argue that, even with other people, we cannot observe subjectivity directly. We have no direct access to it. Rather, we perceive it indirectly in the course of interaction.

Chapter 7 examines how we sense the subjective presence of animals. The discussion combines insights from William James's (1950 [1890], 1961 [1892]) attempts to explore subjective experience with Gene Myers's (1998) research on children's interaction with animals, which builds on Daniel Stern's (1985) work on the emergence of the self in infancy. Although my interests are in adult experience, these works provide insight into pre-linguistic capacities for selfhood. Because several indicators of selfhood appear before the acquisition of language, it is reasonable to look for them in other highly social animals besides humans. Using this framework, the chapter investigates the capacities animals have that allow us to perceive them as having subjectivity. The discussion describes four domains of experience through which we order the world around us. These, in turn, serve as empirical indicators of a "core" self, distinguished as capacities for agency, coherence, affectivity, and self-history. Drawing on data from interviews and observation with animal guardians, I systematically illustrate how dogs and cats manifest these four elements of core selfhood. Animals' core selves become available to us through our interaction, and interacting simultaneously confirms the existence of a core self within us. The feeling of connection is thus not simply anthropomorphizing. Rather, it originates in a match between the core selves of guardians and animals.

The core self goes to work, so to speak, in the capacity for intersubjectivity, by which I mean shared subjective experiences. In Chapter 8, guardians describe instances in which they shared intentions, a focus of attention, or emotions with their companion animals—despite the lack of a shared verbal language. The discussion focuses on play as an activity that evokes all aspects of intersubjectivity and enriches the experience of selfhood for humans and non-human animals alike.

The book concludes by extending the findings to theory and practice. I discuss the theoretical relevance of animal selfhood in an era that is intellectually dismissive of "the self" as a concept. Arguing that the demise-of-the-self critics lack empirical support for their argument, I maintain that, in light of evidence, we *must* theorize the selves of animals. I then argue that, given the weight of the evidence in support of animal selfhood, that status must change our treatment of animals. By first outlining the major positions on animals (that is, welfare and rights), I conclude by saying that the logical—and moral—choice is to support the notion that animals have equal inherent value and must not be treated as property. Most Western societies have long acknowledged that animals can feel pain and thus have an interest in not suffering. Humane-treatment laws are evidence of widespread agreement on our moral obligations not to cause unnecessary suffering. However, we need to justify those obligations by extending equal consideration to animals, for it is illogical to have moral obligations to things. This will mean that we cannot treat animals as property, for their interests in *not* being treated so deserve equal consideration.

Because this research took place in a shelter, the setting controlled the conditions under which people and animals met. In other words, I studied only those who adopted homeless animals rather than, say, those who found strays, who got their animals from other people, or who bought animals from breeders. These other relationships deserve investigation, too. Although the animals in this research came in a variety of colors and sizes and from disparate backgrounds, the people were white and mostly middle to upper-middle class, reflecting the demographics of the setting. This is an interpretive study, and although it is based on empirical research, it is aimed at theory building. It describes what appeared within a sample that was available to me. I hope that it can provide a point of departure for studies that include a greater diversity of participants, both human and non-human.

1 How and Why

Not surprisingly, the species that first achieved the status of domestic companions were those, such as the dog and cat, that were already better adopted to fit this role.

—Peter Messent and James Serpell (1981, 19–20)

This research focuses on our bonds with dogs and cats for several reasons. First, they are numerically the most popular companion animals in the United States, with nearly 60 percent of all households including either or both species (American Veterinary Medical Association 2002).[1] Granted, many households include fish, birds, rabbits, hamsters, reptiles, ferrets, and other animals. However, all of the "specialty" or "exotic" animals taken together occupy only about 10 percent of American households. Second, few other species can share our lives and our homes in the ways that dogs and cats are able to. Although some people have rabbits who sleep with them and birds who know when they are coming home, dogs and cats are uniquely suited to living closely with us. Third, and related to this suitability, few other species have been domesticated for as long as or to the extent that cats and dogs have been. This chapter begins by discussing the "how" and "why" of the domestication of dogs and cats, then examines various explanations of why we continue to want them in our lives.

Domestication can be defined as the process through which the care, diet, and, most important, breeding of a species come under human control. The first animals to become domesticated were canids, members of a family of carnivores that includes thirty-eight species such as the coyote, the domestic dog, and the wolf. Indeed, early dogs probably helped domesticate other species, because their herding and guarding abilities suggest their involvement in the management of grazing animals, such as cows, sheep, goats, and cattle. Scholars disagree on exactly when dogs were domesticated. There are also several views on their ancestry.[2] One names the wolf, *Canis lupus*, as the dog's progenitor. Another argues that dogs are hybrids of wolves and other species of the genus *Canis*, such as coyotes and jackals, and a third view suggests that wild canids, such as the pariah dog of North Africa and Asia and the dingo of Australia, are the domestic dog's wild progenitor.[3] In any case, the domestic dogs who share our homes today exist because of human intervention, which incorporated both cultural and biological processes (see Clutton-Brock 1994, 1995). Biologically, domestication resembles natural evolution. Through selective breeding, humans can—and did—introduce changes in behavior, size, color, ear and tail position, and other features within just a few generations. Culturally, domestication means that a species is "enfolded into the social structure of the human community" (Clutton-Brock 1995, 15). The wolf became a dog, for instance, not just because its physical and behavioral characteristics changed, but also because these changes adapted wolves to material, aesthetic, and ritual purposes in human communities. That adaptation, in turn, may have initiated other changes. For example, Helmut Hemmer (1990) points out that domestication lessens responsiveness to certain kinds of stress. This consequently produces physical changes. Domestic dogs generally have shorter coats than wolves because they do not have to live in the wolf's harsh environment, and the greyhound can run faster and see better than a wolf but may not hear as well. In short, the perceptual worlds of domestic animals are markedly different from those of their wild ancestors. I experienced something of this when my dog Skipper encountered two young foxes at play while he was off his leash in a large field near their den. Once the foxes saw Skipper (I was some distance away), they froze and then ran, but Skipper attempted to initiate play. To the foxes, Skipper was a threat; to Skipper, accustomed as he was to playing with other four-legged visitors to the field, the foxes were potential playmates.

Regardless of how the domestication of dogs occurred, as a species dogs have fared extremely well under domestication. They have adapted marvelously to human society (see Budiansky 1992). Whereas dogs exist throughout the inhabited world, the canids who remained wolves have been eradicated with ferocity.[4] The dog's success stands out because very few species are actually well suited to the process. Francis Galton, a pioneer of modern thinking on domestication (and cousin to Charles Darwin), maintained that candidates for domestication "should be hardy and able to survive with little care and attention. They should have an inherent fondness for humans. They should be comfort loving and useful. They should be gregarious and hence easy to control in groups" (as quoted in Sheldrake 1999, 18). In short, dogs fill the requirements quite nicely.

The best-known account of canine domestication is also the least tenable. It credits dogs' superior hunting abilities as the catalyst for relationships with humans. In this depiction, human hunting parties followed packs of wild dogs. Once the dogs made a kill, humans moved in and took the carcass, leaving scraps for the canids. Over time, this may have evolved into a symbiotic relationship in which predatory canids helped shape human hunting technology. The support for this argument comes from the earliest extant discovery of remains of a domestic dog—clearly not a wolf—from a burial site in what is now Oberkassel, Germany (see Serpell 1988a; Clutton-Brock 1995). The findings date from a cultural period that witnessed the introduction of tiny stone arrows called microliths, which replaced heavy axes for hunting. In this view, the efficiency of microliths would have depended on the use of dogs who could chase and bring down wounded animals. However, another account holds that "the idea of an early hunting symbiosis between men and dogs is a myth" (Messent and Serpell 1981, 8). Dogs bred specifically for hunting with humans are a relatively late development, and hunter–gatherers did not (and still do not) use dogs in the hunt, although dogs often accompanied hunting parties (see Sauer 1952). Instead, commensalism, or shared scavenging, may have led to the domestication of dogs. The catch in this argument is that Mesolithic human settlements "were probably too small and produced too little refuse to provide waste in sufficient quantities to sustain a permanent population of scavenging wolves" (Messent and Serpell 1981, 9).

A third view, and one I find convincing, suggests that the domestication of the dog was "free of utilitarian considerations" (Messent and Serpell 1981, 10). The historical record shows that early humans tamed animals of all sorts. Useful canine skills in hunting, guarding, and herding may have cemented the relationship between humans and dogs, but they probably did not initiate it. Instead, several biological and behavioral factors predisposed dogs to fit easily within human groups. For one, canids have a long primary socialization period. There are several months during which, given sufficient contact, puppies can form attachments to humans. In contrast, animals such as cows and horses are precocial at birth. On their feet immediately, they become adults much sooner than dogs do. Although they can bond emotionally to humans, they do not incorporate humans into their social groups. Dogs, in contrast, relate to humans as litter mates and pack members. As Constance Perin (1981, 80) explains, "The human family provides a parallel to the sort of group dogs are equipped to relate to. In the 'good family dog' we recognize that biological basis for the two species coming together." Another factor that predisposes dogs to human companionship is their intense interest in play, a topic I explore in later chapters. Because dog–human play is non-competitive, it can be enjoyed by people of any age, offering opportunities to enhance the interspecies bond. In addition, dogs, like humans but unlike wild canids, are diurnal.[5] Dogs' activity cycles mean that they are alert when their people are also awake. Moreover, because dogs are particular and habitual about where they eliminate, they can be house-trained more easily than many other animals. Their physical size may also be a factor: Even the largest dog is still smaller than an adult human and can live in human homes more easily than can, say, horses, giraffes, or elephants.

For these reasons, and probably others, dogs were predisposed to the role of companion animal. Once they had filled that niche, no alternative species were necessary. Although numerous other animals have become companions to humans, none has done so with the success of dogs.

"Nor have they wanted to," I imagine my cats saying in unison. The origins of the domestic cat are more difficult to trace than are those of the dog. Although the dog's morphology differs significantly from that of its wolf ancestors, the cat differs only slightly from its presumed ancestor, the North African wild cat, *Felis sylvestris libyca.*[6] Moreover, "domestication" hardly seems an accurate term for portraying the feline scenario, because that scenario involved "deification" as well. Both processes apparently began in Egypt around 5,000 years before the common era, as large agrarian societies were emerging (see Clutton-Brock 1981). Whereas the domestication of dogs occurred *prior to* the development of agriculture, the domestication of cats probably occurred *because* of it. The Egyptians valued cats because they hunted the rodents that threatened stores of grain, the basis of Egypt's economy. Cats gained not only esteem; they also, ultimately, gained the status of the deity named Bast, the goddess of joy, fertility, and motherhood (see Berghler 1989; Siegal 1989). This marked the beginning of the cat's association with the feminine, which in other cultures would bring a stigma and even a death sentence. In ancient Egypt, however, laws protected cats; temples honored them; and art paid tribute to them. The cat's incorporation into Egyptian family life began what one author has called "the age of glory in the checkered history of the domestic cat" (Siegal 1989, 4). A cat's death meant that the family went into mourning, shaving off their eyebrows as a sign of their loss. If the family had wealth, the cat received an elaborate funeral; archaeological digs have revealed numerous remains of mummified cats. Egyptian laws prohibited exporting cats, but cats eventually reached Europe by way of Greece around the sixth century B.C.E. (Berghler 1989; Málek 1993). Archaeological evidence reveals the presence of domestic cats in Britain from the middle of the fourth century and throughout Europe by the tenth century (see Serpell 1988b). Cats did not arrive in North America until much later, when they accompanied white, European settlers.[7]

Although originally drawn to humans because we offered a steady food supply, cats maintained their independent characteristics. Like the association with the feminine, this worked against cats, who were considered not only "sexually charged" but also "dangerous, egotistical, and cruel" until the late nineteenth century, when their image was "rehabilitated" sufficiently for them to become acceptable house pets like the dog (Kete 1994, 116; see also Ritvo 1988). The cat's incorporation into middle-class homes signaled a change in attitudes whereby the allegedly excessive independence and sexuality of the cat became less threatening. However, even today cats retain vestiges of stigma. For example, I know of no canine equivalents to such best-selling books as *A Hundred and One Uses for a Dead Cat* (Bond 1981).

One author claims that cats may be *domestic,* but they have never been *domesticated* to the extent that dogs are (see Leyhausen 1979). As a species, cats have not responded to human efforts at selective breeding as enthusiastically as dogs have. Although there are more than 400 dog breeds, there are fewer than 50 breeds of cats.[8] Unlike dog breeds, which vary widely in size, temperament, and other physical and behavioral characteristics, cat breeds differ primarily in coat color and length.[9] Like dogs, cats are pre-adapted in a number of ways to live with humans. They have a long primary socialization period that allows kittens to bond with humans. Cats sleep a considerable amount—up to twenty hours a day—but they are crepuscular, having periods of high activity at morning and evening, when most humans can enjoy their company. Granted, anyone who has lived with cats knows that their idea of "morning" is often much earlier than many people would like to start their day. Still, feline rhythms coincide more or less with those of humans. In addition, cats will readily use litter boxes and can adapt to life in even the smallest apartment. Humans can easily participate in cat play, as well. However, some factors make cats' incorporation into human life distinct from dogs', and somewhat incomplete. As a species, cats are not gregarious (although there are certainly individual exceptions). Cats are territorial, but they also love comfort; thus, they formed symbiotic relationships with humans while retaining qualities of the solitary hunter. The ubiquity of present-day feral-cat colonies attests to the ease with which they revert to life without humans.

In sum, although human societies have incorporated animals of all kinds, certain members of the Canidae and Felidae families were predisposed to coexist with us. The factors that made them so are not simply anthropomorphic projections. Rather, they are behavioral and biological traits that exist independent of our perception of them. For example, the feline instinct to seek loose material on which to urinate and defecate existed prior to the invention of cat litter, not because of it. The canine instinct to bark at threats to the group existed independent of the human desire to guard property. Cats and dogs did not learn these behaviors from humans. They are instinctual, and they appear at developmentally appropriate times. Granted, humans manipulated instinctive traits in domestication, but the traits existed before that.

Despite the suitability of dogs and cats for life with humans, or perhaps because of it, our relationships with them are unique. We humans are, for example, the only species that habitually and intentionally adopts other species into our midst. Although the relationship is mutually beneficial, its existence raises a number of questions. Why, for example, do we invite a carnivorous predator such as the dog into our homes? Why extend the invitation to cats, with their razor-sharp claws and tendency to regurgitate the fur they swallow while grooming?[10] The relative ease with which dogs and cats adapted to life with humans explains the start of our mutual relationship. I turn now to some of the accounts of why we continue to want to have animals around us.

WHY HAVE PETS?

The Deficiency Argument

One explanation of the appeal of dogs and cats maintains that our relationships with them are surrogates for the relationships that we should have with other people (see Shepard 1978, 1996). In this view, human–animal relationships are distorted and deficient substitutes for human–human relationships. I call this the deficiency argument, because it assumes that people who enjoy the company of animals lack the qualities or skills that would allow them to enjoy human company. Other targets of the deficiency argument include environmentalists and animal-rights activists, who are derided with claims that they would rather save a tree or a lab rat than a human being.

The deficiency argument has a long history. As I explain in the next chapter, the ancient association of animals with perversity endured well into the modern era, representing Western anxieties about the boundary between humans and animals. It took its most violent form in the witch hunts, in which companionship with animals alone was evidence of consorting with the devil, and the animal friends of the accused went to death with him—or, more likely, with her. Contemporary versions of the deficiency argument are favorites of the media, making it easy to get a distorted view of human–animal relationships. One does not have to look far to find accounts of people—usually women who live alone who are so concerned about saving animals that they become "hoarders," as they are known in animal-welfare circles, accumulating far more creatures than their resources allow them to care for (see Arluke et al. 2002). Other stories emphasize the extremes—financial and otherwise—to which people will go for animals because they have no other family members. Then there are accounts such as the one of the woman whose husband left her because she loved her dog more than she loved him.[11] Granted, some people have remarkable relationships with their animals. Nevertheless, stories that focus on the extremes distort what we know about the majority. Just as stories of alcoholics and anorexics reveal little about typical, moderate drinking and eating, stories of hoarders and eccentrics teach us nothing about average relationships with dogs and cats.

In addition to its sensationalism, the deficiency argument has two fatal flaws. First, no extant study reveals evidence of any qualities, skills, or lack thereof that predispose certain people toward companionship with animals. If animals were substituting for relationships with humans, then we could expect to find evidence that "animal people" differ on some significant psychological indicator from "non-animal people." Psychologists and others have tried to demonstrate this, to no avail. The most comprehensive review of personality research on animal caretakers failed to find definitive differences between those who had animals and those who did not—or, for that matter, between people who had dogs as companions and those who had cats (see Podberscek and Gosling 2000).[12] Most studies indicate that people who enjoy animal companionship are, on average, more or less like everyone else. One study suggests that people who do not have companion animals do not avoid having them because they dislike animals (Guttman 1981). Instead, people who do not have animals display somewhat stronger tendencies to avoid permanent ties than animal caretakers do. They also tend to place more importance on cleanliness in their home environment. The same study shows that animal caretakers more often mention the benefits of companionship as a reason that they have animals. They report feeling alone without an animal. Once one is accustomed to animal companionship, the relationship is not easy to do without. In my research, I found this to be the case. For example, many people spoke of the empty feeling in the house after the death of a pet and looked forward to having a dog's or cat's company again. In addition, people who could not have animal companions because of lifestyle issues, such as frequent travel or a landlord's prohibitions, often said they looked forward to the time that they could again have dogs and cats. This is consistent with research demonstrating that the most significant difference between people who have animals and those who do not seems to be that those who have animals usually had them during childhood (Poresky et al. 1988). Overall, however, there are probably more differences within the category of "animal people" than there are between "animal people" and "non-animal people."[13]

The second flaw in the deficiency argument is this: If animals offer substitutes for human relationships, then we could expect to find the highest frequencies of pet ownership among single people. Instead, single-member households are the *least* likely to include companion animals (American Veterinary Medical Association 2002). Animal companionship is highest in households with parents and children. Similarly, no evidence indicates that relationships with animals detract from or interfere with relationships with other people. Instead, data suggest that animals serve as "social facilitators" (see Messent 1983; Robins et al.

1991; Sanders 1999). They augment our relationships with other humans, and this is especially the case for dogs. For example, people accompanied by dogs in public places have more frequent and longer interactions with others than do those without dogs (Messent 1983). In addition, dogs allow people to violate the tenet of "civil inattention" that Americans observe most of the time (Robins et al. 1991). "Civil inattention" is Erving Goffman's (1963) term for the way we notice the presence of another person, say, across from us on a bus or in the next line at the grocery checkout but typically avoid making eye contact or otherwise encouraging further interaction. We look into the middle distance rather than stare. Dogs transform their human companions into what Goffman calls "open persons," amenable to greetings and conversations. A dog can also serve as what Carol Brooks Gardner (1980) calls a "badge." Dogs announce that their human caretakers are "dog people," which makes them susceptible to interaction with other "dog people."[14] A personal anecdote makes this point nicely. About a year after adopting Skipper and purchasing a condominium, I attended a meeting of the homeowners' association. I sat next to a neighbor who had lived in the complex for about four years. As we waited for the meeting to start, about a dozen people said hello to me or stopped to chat, and one invited me to her place for a drink after the meeting. This dumbfounded my neighbor, who said that in four years she had come to know only me and one other person. I explained that I knew all these people through our dogs, who "introduced" us to more of our neighbors than we would ever have met on our own. We saw one another regularly on walks, sometimes twice a day. We learned our dogs' names long before we learned one another's.[15] The same holds for cats but to a lesser extent, because they do not typically accompany their human friends in public. Nevertheless, "cat badges" allow for similar violation of civil inattention. For example, people have struck up more than a few conversations with me because they have noticed that my socks have cats on them or that I am wearing cat earrings.

Clearly, the deficiency argument fails to show that animals substitute for human relationships. There is, however, an additional point worth mentioning. Ample research shows that animal companionship is good for people physically, mentally, emotionally, and otherwise (see Fogle 1981; Siegel 1993; Beck and Katcher 1996; Wilson and Turner 1998; Podberscek et al. 2000). Animals can be therapeutic in numerous ways, even when their role is not explicitly so. Is this not evidence of human deficiency remedied by animal companionship? I argue that it is not. In these cases, animals are not substituting for something that humans could or should get from other humans. They are providing something unique to animal companionship. The relationship is quite different. As Aaron Katcher (1981, 50) argues, "Pets are not substitutes for human contact but offer a kind of relationship that other human beings do not provide." They have an innate and highly therapeutic ability to accept us as we are. Regardless of our mood when we come home at the end of a day, they are there "stirring up the dead air in a room," as Thoreau is said to have put it (see Perin 1981, 79).

In sum, the deficiency argument points out the need for empirically grounded, theoretically driven work on relationships between humans and non-human animals. As Gene Myers (1998, 64) contends, "Until we have a clear account of human–animal relationships, we may be bound to look at them as distorted human–human relationships."

The Affluence Argument

Another explanation equates animal companionship with economic prosperity. More accurately, the affluence argument implies that it is a waste of money to feed animals when we should be caring for other people. Perhaps the first instance of this argument appeared in the writings of Plutarch. It gained strength in the Middle Ages when church authorities used it to prohibit nuns and monks from having companion animals.[16] Harriet Ritvo (1987) found ample documentary evidence of the affluence argument in nineteenth-century England, and it evidently still carries considerable ideological weight, because in the United States, recipients of food stamps may not use them to purchase dog or cat food. In this view, the fear seems to be that the poor will deprive themselves—or, worse yet, their children—so their animals can eat. Although examples of people who go without food to feed their animals probably exist, these hardly represent the norm. In the course of this research, I encountered instances in which people surrendered animals because they could not afford to keep them. Alternatively, poorly cared for animals often came from neighborhoods where the homes cost more than a million dollars.

The idea that money spent on animals should rightly go to feeding the poor trivializes our obligation to companion animals. When we domesticated certain animals, we took on responsibility for their care (see Rollin 1992; Beck and Katcher 1996).[17] Moreover, it does not logically follow that a commitment to animals leads to the neglect of human social concerns. The belief that concern for animals detracts from concern for human suffering has been labeled the "displacement thesis," in that it represents displaced compassion for other people (Finsen and Finsen 1994, 26–30). On the contrary, people who are involved in animal welfare or rights are often active in or supportive of other social causes on behalf of humans (see Nibert 1994). This has long been the case. Many efforts on behalf of animals arose contemporaneously with those on behalf of humans, such as reform in education, prisons, and mental hospitals; abolition; and suffrage (see Turner 1980). Several of the charter members of the American Society for the Prevention of Cruelty to Animals (ASPCA) were well-known abolitionists, as were the founders of the SPCA in Britain. Henry Bergh, the ASPCA's founder, also established the first Society for the Prevention of Cruelty to Children when, in 1874, a social worker named Etta Wheeler approached him with a difficult case. She had been trying to have young Mary Ellen McCormack removed from her terribly abusive foster parents. Authorities would not, and legally could not, intrude into the family, even to protect the child. Wheeler went to Bergh, who won release of the child and successfully prosecuted the parents (see Coleman 1924; Finsen and Finsen 1994). Other examples abound. Frances Power Cobbe, the British founder of the antivivisectionist Victoria Street Society, was a suffragist. In the 1970s, many of those who became active in the animal-rights movement were veterans of civil-rights activism. Susan Sperling's book *Animal Liberators* (1988, 111) provides accounts of in-depth interviews with animal-rights activists, most of whom see themselves as "fighting a system that abuses women, minorities, and animals." Granted, some animal-rights activists have injured people and destroyed property in the course of their efforts (see Jasper and Nelkin 1992). Moreover, the Nazis sponsored a strong program of animal protection alongside horrific disregard for human life (see Arluke and Sax 1992; Arluke and Sanders 1996). These anomalies aside, some scholars have suggested an "extension thesis," which, in contrast to the "displacement thesis," holds that "those who devote themselves to the welfare on one exploited group (whether human or animal) in many cases extend concern to other groups as well" (Finsen and Finsen 1994, 28). Nibert's research suggests a relationship between support for animal rights and support for issues such as gun control and rights for women, homosexuals, and people of color. Those who opposed animal rights were more likely to

> favor easy access to guns, to oppose abortion rights, to exhibit racial prejudice, to be more approving of interpersonal violence, to blame the victims of rape and to exhibit prejudice against homosexuals, and less likely to give people with different sexual orientations a right to free speech. (Nibert 1994, 122)

From any angle, the affluence argument fails to convince. Pets do not exist only in affluent societies; nor do they eat food that rightly belongs to poor human beings. Concern for animals does not amount to neglect of human needs. In light of the evidence against it, the affluence argument turns out to be ideological, resting on the anthropocentric claim that humans matter more than animals.

Here is another flaw in the affluence argument: Although it is true that pet keeping—at least, in England and the United States—grew apace with economic advances, it is highly unlikely that economic security caused the increase in pets. Rates of pet keeping skyrocketed during the late nineteenth century, but the potential contributing factors are too numerous to pin to a single cause. Although I examine these more closely in the next chapter, I will mention some here. First, some credit must go to Darwin. His notion of a kinship among species lessened fears of contamination by association with animals and increased curiosity about them. Moreover, as machinery gradually replaced animal labor, animals began to symbolize an older, simpler way of life (see Thomas 1983). At the same time, advances in veterinary medicine, animal husbandry, and weapons technology made "those who had to deal with animals less vulnerable to natural caprice" (Ritvo 1988, 20). Nature in general was no longer a constant menace, so human beings could view it with more affection. It could even serve as an antidote to hectic, modern life. The Romantics could claim to feel at home in the wilderness, where they escaped the modern "progress" that allegedly stifled human potential (see Noske 1997). In addition, during the mid- to late nineteenth century, pet keeping, long restricted to elites, became less so. At the same time, the establishment of the first humane shelters provided "an opportunity for many to secure valued pets, a privilege appreciated by hundreds who could not otherwise secure them" (Coleman 1924, 210). In short, increasing economic security surely helped to promote the popularity of animals, but so many factors coalesced at the same time that it is unlikely to have been the sole cause.

The Dominance Argument

Another explanation maintains that companion animals allow us to manifest power over nature. From this perspective, pets are one example among many, including gardens, aquaria, fountains, bonsai trees, and topiary art. The best-known proponent of this view is Yi-Fu Tuan, from whose book *Dominance and Affection: The Making of Pets* (1984) I drew the name for this argument. Tuan begins with the unquestionable claim that "any attempt to account for human reality seems to call for an understanding of the nature of power" (Tuan 1984, 1). He then acknowledges that the concept of power offers only a "partial and distorted" picture, for as much as people attempt to dominate, they also cooperate with and care for one another. Affection, therefore, is also important for understanding "the day-to-day maintenance of the world" (Tuan 1984, 1). Instead of posing affection as the opposite of dominance, Tuan poses it as its anodyne, its less-offensive version, or "dominance with a human face" (Tuan 1984, 2). "Dominance may be cruel and exploitative," he writes, "with no hint of affection in it. What it produces is the victim. On the other hand, dominance may be combined with affection, and what it produces is *the pet*" (Tuan 1984, 2; emphasis added).

At first, I found Tuan's ideas disturbing. I took offense at the suggestion that my dogs and cats, who live comfortably and seem to want for nothing, were manifestations of my need to dominate other beings. I think that most people who truly care for animals would probably find the dominance argument unsettling, because it implies abuse. However, Tuan claims that, although power is subject to abuse, it "is not *inevitably* abused" (Tuan 1984, 176; emphasis added). He argues that inequalities of power make true affection possible. In human relationships, this inequality can produce the affectionate form of dominance. He illustrates his point with the example of the intimacy of marriage. He describes the "temporary bonds of inequality," such as those that appear during sickness, when one partner depends on the other for care (Tuan 1984, 163). Each partner in a marriage knows that there will be occasions when he or she will be dependent, and this vulnerability makes intimacy possible. Two equal individuals can never enjoy the level of intimacy and affection that is possible in marriage, Tuan argues. Exercised in this way, power is, as Tuan puts it, "creative attention," which is also known as love (Tuan 1984, 176).

With companion animals, the situation differs slightly. Our relationships with them are necessarily unequal. They depend on us to give them food, water, and even to allow them to relieve themselves. In addition, the guardian—at least, the responsible one—will exert power over the animal in training, vaccinating, sterilizing. He or she will also exert power in the many daily instances in which the dog or cat wants to do something—go outside, come inside, bark at the mail carrier, scratch at the upholstery—and the guardian must control the animal's behavior. Much of this control is for the safety of the animal and, in any case, it is an unavoidable aspect of the relationship. In Tuan's argument, the pleasure of animals' company originates in our ability to be "masters" over them. We manipulate their behavior in ways that are not natural and that other humans would not tolerate. We give them childish or silly names. We tease them. Because the pleasure of having a pet comes largely from the animal's obedience, punishment, abandonment, or neglect await pets who do not obey. Then, when they become old, inconvenient, or tiresome, we get rid of them. The inevitable conclusion, according to Tuan, is that we use animals. "Whether we use [them] for economic or playful and aesthetic ends, we *use* them," he writes, "we do not attend to them for their own good, except in fables" (Tuan 1984, 176; emphasis in the original).

Tuan makes many valid points. I have encountered numerous examples of the inhumane things that people will do to animals to "sanitize" them of their natural, animal qualities. They will declaw cats, for instance, amputating the end bones of the toes along with removing the claws so that they do not damage furniture. They will sever the vocal chords of dogs so that they cannot bark.[18] They will use choke collars and other forceful means to get dogs to "mind." They will attempt to breed hypoallergenic cats. Then, when the animals still refuse to fit the ever-changing human lifestyle, they will relinquish them to shelters or simply abandon them. In short, we do not need to look far to find support for the dominance argument.

However, like the affluence argument, the dominance argument describes only one type of relationship with animals: that of "owner" and "pet" (see Serpell 1986). The image that comes to my mind is that of "Tricky Woo," the overfed Pekingese from James Herriot's books. Tricky was the paradigmatic "pet." He lived with Mrs. Pumphrey, a rich widow who pampered him with the richest of food, which inevitably brought on digestive troubles that required frequent house calls from Herriot, the vet. In addition, Mrs. Pumphrey summoned Herriot for countless other emergencies, most of which were the products of her imagination and boredom. Afterward, Tricky "wrote" Herriot thank-you notes and included gifts of kippered herring or good cigars. Clearly, Mrs. Pumphrey was accustomed to keeping pets.

Such relationships with animals do exist. Animals can be playthings. However, that is neither the only nor the most important aspect of our relationships with non-human animals. They can be friends, eyes, ears, and more. Moreover, the practices connoted by the term "pet" should not be lumped together with the more humane and cooperative practices that constitute relationships with "companion animals." Although I discuss this at length in the next chapter, for now I will say that the term signifies an effort to appreciate animals for what they *are*, not for what they *could be* if only they were not so much like "animals." Whereas a pet must please and entertain a human "master," a companion animal has a guardian or caretaker who acknowledges the animal as one whose ways of being in the world are radically different but still worthy of respect. A guardian who sets out to train a companion dog, for example, would begin by understanding something about how dogs learn and how they establish pack hierarchy even in human homes. The guardian would learn the importance of simple acts such as going through doors before a dog in training or eating before feeding the dog. Such actions communicate to the dog that the guardian is the leader of the pack, and they do so in language the dog can understand: *Leaders go first.* The owner or "master" of a pet, in contrast, would either turn the dog over to someone else for training or do it badly, yelling at the dog for urinating on the rug hours after the accident occurred, thereby sending the dog a confusing signal in a foreign language that tells him nothing about how to do things right.

I hesitate to agree that the same human tendencies expressed in "pet keeping" also describe all types of relationships with animals. The practices associated with the keeping of pets may well manifest a human need to dominate other beings, but I am not convinced that the same need, even in the form of an anodyne, also explains practices that constitute alternatives to pet keeping. To be sure, our animal companions did not freely choose to be with us. Although their utter dependence on their human guardians indeed creates a fundamental inequality, this does not necessarily translate into domination. After all, children depend on their parents, too, but parenting has to do with something other than domination. In both cases, the responsibility of caring for and nurturing another creature offers pleasures and rewards that have nothing to do with domination.

I am wary of reducing all behavior to one cause, especially one that seems as tautological as power. If abusing animals is evidence of power and kindness to animals is evidence of power, as well, that leaves no alternatives. Power is everywhere, and everything is power. If this is the case, then accepting power as the orchestrating force is a matter of faith. Although I acknowledge that power is an important force to consider in all social relations, I am not convinced that it explains everything. The relationships between people and animals are too varied and too fluid to attribute to a single causal factor.

The Biophilia Hypothesis

Perhaps people have a natural connection to animals—it sometimes seems that way—and *this* accounts for the widespread appeal of dogs and cats. This is the gist of another explanation, not just of the appeal of pets, but of relations between humans and nature. In 1984, the biologist Edward O. Wilson published *Biophilia,* in which he proposed that humans have an "innately emotional affiliation" with other living organisms—"innate" in the sense that it is "hereditary and hence part of ultimate human nature" (see Wilson 1993, 32). In other words, Wilson speculated that meaningful human existence depends so heavily on our relationship to the natural world that it has a genetic and evolutionary basis. More precisely, it has a *biocultural* basis. Our genes and culture co-evolve over time; genes influence certain behavior that enhances survival and reproductive fitness; and as language and culture develop, they simultaneously encompass the behavior that has spread throughout the population by natural selection. Wilson illustrates the process with an account of the relationship that humans have with snakes. Many people have an aversion to snakes, even with little exposure to them. Yet snakes also fascinate us; we view them in captivity with a kind of squeamish attraction, and many cultures use them in religious symbolism. People dream of snakes more often than any other animal, Wilson claims. Our primate ancestors also have similar aversion-and-attraction responses to snakes. In nature, poisonous snakes pose a significant threat to primates and, on seeing one, an ape or monkey will vocally warn the others in the group. Instead of running from the snake, however, the group will follow it until it leaves the area, demonstrating a fascination with snakes that parallels that of humans. According to Wilson, natural selection encoded our primate ancestors' need to avoid snakes, combined with their frequent and inevitable encounters with them, as a hereditary fear and fascination. As human cultures evolved, they carried this hereditary response with them, and it manifested itself in mythology, stories, artwork, and dreams. Thus, a biological imperative—the need to avoid snakes—simultaneously produced behavioral *and* cultural responses. Different responses, such as affiliation with animals, could have emerged by the same process but under different "selection pressures and with the involvement of different gene ensembles and brain circuitry" (Wilson 1993, 34).

On the face of things, there seems to be ample evidence of the "innate" attraction to nature and animals. People value homes with views of the forest, mountains, or water. They decorate the interiors with plants that bring nature indoors. The art that hangs on the walls is more likely to depict a landscape than any other subject (see Halle 1993). People enjoy watching animals. The number of people in the United States who visit zoos exceeds the number of those who attend all the major professional sporting events combined. Moreover, captive environments now attempt to simulate natural habitats, and animals in many major zoos no longer live their lives in solitary cages. Birdwatching, even the backyard variety, has become sufficiently popular to support franchises that sell feeders, seeds, books, binoculars, and related items. For those who can afford them, activities such as whale watching offer opportunities to see animals at close range. Even at home, people have access to a wide range of nature and wildlife documentaries on television. Simultaneously, advances in videography and other technology have provided viewers with ever closer glimpses of animals in their habitats. Interest in animals generated Animal Planet, a cable-television channel devoted entirely to animal programs that range from Steve Irwin's irrepressible *Crocodile Hunter* to investigative reporting of animal-cruelty cases. The fascination with animals begins at an early age. Children seem to have a natural rapport with animals. They pretend to be animals; they talk with animals; and animal characters populate their toys and cartoons (see Myers 1998; Melson 2001). Added to these examples are the growing numbers of companion animals. Surely these represent an innate human tendency to affiliate with nature.

Or do they? The biophilia hypothesis, too, contains several flaws. First, it is a-historical. It mistakes late–twentieth-century developments for long-term, universal human tendencies. For example, before Rachel Carson's *Silent Spring* (1962) was published, concern for the environment was on few people's minds. For much of modern history, people considered forests "dreadful," "gloomy" places (see Thomas 1983, 194) in which only eccentrics could find beauty. Animals, long considered "brutes" and "beasts," existed to serve human purposes. Today, so many people in contemporary Western societies live so closely with animals that it is helpful to remind ourselves that we have not always done so—or wanted to. As Arluke and Sanders (1996, 191; see also Franklin 1999) explain, "How we think about animals, as well as ourselves, is bound to change as society itself changes." Even a generation or two ago, activities such as whale watching were unheard of, and houseplants consisted mostly of the African violets grown by grandmothers. Granted, the wealthy could always build homes that allowed them to appreciate the natural world, but the overall trend was to cut down trees to make building easier, drain wetlands to "reclaim" land, and divert rivers for hydroelectric power. Only somewhat recently have the negative consequences of these attempts to "manage" nature become widely apparent. Even so, decisions about how to balance human interests with conservation remain highly contested.[19]

If it were in humans' genetic interests to preserve and appreciate nature, then what could explain our apparent dedication to its destruction? A proponent of biophilia would argue, as Stephen Kellert (1993, 42) has, that "even the tendency to avoid, reject, and, at times, destroy elements of the natural world can be viewed as an extension of an innate need to relate deeply and intimately with the vast spectrum of life about us." Herein lay another problem with the notion of biophilia: In explaining both the preservation of nature and its destruction as expressions of the same ultimate end, it is tautological. In addition, it requires accepting the existence of some prevailing entity that has "needs" that must be satisfied. To accept biophilia, for example, I must first accept the primacy of the hereditary "needs" of the human species. As was the case with the primacy of power in the dominance argument, this is a matter of some faith. Granted, it makes our ethical obligation to nature a biological imperative (see Kellert 1993). However, in making it so, biophilia runs roughshod over the potential and various meanings that individuals give to their relationships with nature and other animals. Perhaps we *do* have a genetic tendency to take an interest in animals, but if that is so, our explanations still need to go further. As Myers puts it, "Even if we have biologically programmed 'learning rules' for biophilia, we need to account for this potential in the context of our full human capacities" (Myers 1998, 45).

It would be satisfying to understand why we have close relationships with animals, particularly dogs and cats. In many ways, they are such unlikely friendships. By the end of this book, I will have pieced together an answer. The pieces I lay out along the way will show how everyday relationships with animals occur—in lived experience, at this time—rather than attributing them to the playing out of some innate "need" or "force." The meaning of animals changes across history and across individual lives, revealing not only what we think about animals but also what we think about ourselves—as a culture and as individuals. As Adrian Franklin (1999, 53) maintains, "People may have always liked animals but history teaches us that they like them in profoundly different ways under different historical conditions." The next chapter takes that statement seriously and explores its implications.

2

Them and Us

At various times throughout history, in comparing animals to humans, various species have been regarded in different ways that resulted in great variation in the manner in which they were treated. Notions about the capacities and powers of nonhumans have run the gamut all the way from animals as possessing greater powers and capacities than people and therefore being viewed as gods, to being categorized as totally different in every detail, hence having nothing at all in common with our species. Attributions of superiority and inferiority virtually always accompany the designations. And implications usually follow that involve the appropriateness of exploitation being dependent upon differences.

—ELIZABETH ATWOOD LAWRENCE (1995, 75)

There can be no single answer to the question of why people form relationships with dogs and cats because our relationships with them have not been of a universal, standard type that could generate a once-and-for-all explanation. Although people have lived with dogs and cats for ages, the meaning of doing so has varied significantly over time. Living with a wild animal one has captured and tamed differs from true domestication, and in neither case is the animal necessarily a pet. A pet has surpassed categorization as an animal. Animals, in contrast, are often nameless, which makes it easier for us to eat them or experiment on them. Moreover, much of the behavior that repulses or frightens humans when shown by animals is tolerable and even endearing in one's pet. Finally, the term "companion animal" restores some of the "animal" dignity to the "pet."

Each way of understanding, describing, and treating animals is freighted with history. Each era has socially constructed animals and human relationships with them. I am using the term "social construction" in a particularly cautious way—for the most part to refer to Peter Berger and Thomas Luckmann's (1967) notion that certain attitudes and beliefs are "co-constitutive" of a social group's perception of reality. In this chapter and the next one, for example, I offer a selective, historical examination of attitudes, beliefs, and social contexts that produced the categories "animal," "pet," and "companion animal." In saying that these categories were socially constructed, I am pointing out that there is no a priori distinction between a "pet" and an "animal." It is a matter of convention that Americans will eat cows but not poodles.[1] The norms and linguistic practices that allow us to distinguish animals from humans or animals from pets are not given in any absolute sense. They are best described as social constructions, and the human–animal boundary may well be one of the first. The boundary is not "natural"; it is a vehicle for particular human goals and conflicts. Its power lies in the way it is experienced as objective reality.

Having said this, I want to take care to clarify the limits of my use of the term "social construction." I do not mean to say that individual animals themselves are social constructions. As I discussed in Chapter 1, many canine and feline behaviors exist independent of human ideas of them. Nevertheless, we experience individual animals through "complex and sedimented layerings" of categories, some of which are socially constructed (Shapiro 1990, 193). We experience particular dogs or cats, for example, as (among other things) representatives of species ("the dog"; "the cat"), as creatures governed by instinct ("animal"), and, depending on our attitude, as compliant, living toys ("pet"), or four-legged friends ("companion"). Our bonds with individual animals draw on all of these social constructions, especially because we humans have bred animals, particularly dogs, to bring these constructions to life.

Fortunately, our bonds with individual animals involve more than social constructions. As we will see in later chapters, they draw on lived histories and on unique selves. Our relationships are not limited to social constructions, but they are not independent of them, either. To understand everyday lived experience with animals, it will help to understand some of the social constructions that partly inform that experience.

ANTHROPOCENTRISM

Constructing the Category of "Animals"

The category of "animals," as distinct from and inferior to humans, probably appeared with the transition from hunting and gathering to agriculture. Evidence suggests that preliterate peoples lived with nature in a relationship of oneness and respect (see Ingold 1994; Schwabe 1994; Noske 1997). Although they could surely distinguish themselves from animals, there is no evidence that they saw themselves as superior to the creatures around them. They used animals' bodies to meet the material necessities of living, but they also used animals, as beings, to meet spiritual needs. As John Berger (1980, 2) puts it, "To suppose that animals first entered the human imagination as meat or leather or horn is to project a 19th century attitude backwards across the millennia. Animals first entered the imagination as messengers and promises." It is not simply romanticizing to say that many, if not most, preliterate peoples considered animals superior to humans, having magical, even divine powers. Indeed, animals were the first symbols. To offer just two of countless examples, animals represent eight of the twelve astrological signs, and numerous creation myths depict animals carrying the Earth. Preliterate groups saw that animals were like humans in many ways, but also different enough to be able to explain and accomplish things that humans could not. As Berger argues:

> Animals interceded between man and their origin because they were both like and unlike man. Animals came from over the horizon. They belonged *there* and *here*. Likewise they were mortal and immortal. An animal's blood flowed like human blood, but its species was undying and each lion was Lion, each ox was Ox.

> This—maybe the first existential dualism—was reflected in the treatment of animals. They were subjected *and* worshipped, bred *and* sacrificed. (Berger 1980, 4–5; emphasis in the original)

The elements of worship and of a related spiritual connection to animals dropped out of the equation when the means of production changed in human societies. As the anthropologist (and veterinarian) Elizabeth Lawrence (1986, 46) explains, "It is impossible to overestimate the importance of mankind's change from hunter-gatherer to domesticator of plants and animals." Hunter–gatherers took what they needed to survive, but their very survival meant that they could not overexploit the environment on which they depended. In contrast, the transition to farming required both an intimacy with the natural world and a conquering attitude toward it. The farmer works in opposition to nature by eliminating undesirable plants and animals, consequently labeling them "weeds" and "pests." The farmer also manipulates water and the reproduction of crops. Consequently, the transition to farming required "new ideologies . . . that absolved farming people from blame and enabled them to continue their remorseless programme of expansion and subjugation with a clear conscience" (Serpell 1986, 218). Groups with ideologies that distanced them from nature became the most prosperous. The success of settled, agricultural civilizations required an attitude of domination, justified through beliefs that animals were not only "other," but also inferior to humans (see Thomas 1983; Tuan 1984; Franklin 1999). This is the sense in which the human–animal divide is a social construction: It is neither natural nor inevitable; it is the product of the power humans exerted over other creatures. "Progress" required human communities to define the natural world and the non-human animals within it "as fundamentally different and ontologically separate" from their own (Wolch 1998, 121). It produced not only difference but also inequality, for in the case of non-human animals, "different" meant "inferior." Thus, the ideology of anthropocentrism, which placed humans at the center of creation, gradually supplanted the sense of respect for the rest of nature.

The monotheistic religions of Judaism, Islam, and Christianity all justify a strong form of anthropocentrism known as "dominionism," or a God-given right to rule over nature.[2] Here, for example, is the famous passage from Genesis 1:28:

> And God blessed them, and God said unto them, Be fruitful, and multiply, and replenish the earth, and subdue it; and have dominion over the fish of the sea, and over the fowl of the air, and over every living thing that moveth upon the earth.

The directive to "have dominion" was one for which humans, as a species, were woefully ill equipped. Therefore, dominion required the help of animals who had the skills and strength that humans lacked. In allowing certain species into close, working relationships with humans, dominionism gave some animals a special status along the borderlands of the human–animal divide, especially those who assisted humans in their quest for dominion. For example, in declaring, "You shall not muzzle the ox while he is threshing," Deuteronomy 25:4 directs farmers to allow beasts of burden to partake of the fruits of their labor. Likewise, the commandment to observe the Sabbath also dictates that neither "your ox or your donkey or any of your cattle" should work, as well (Deuteronomy 5:14).[3]

Because animals played crucial roles in society, debates arose over their status and appropriate treatment. More accurately, the debates arose because animals were so like us in so many ways. As Berger (1980, 5) puts it, "The parallelism of their similar/dissimilar lives allowed animals to provoke some of the first questions and offer answers." While ancient thinkers struggled with questions of morality and justice, they simultaneously attempted to determine who possessed the requisite mental or spiritual capabilities to be moral and just. Along the way, numerous views of animals were in circulation (see Sorabji 1993). For example, the Pythagoreans and Platonists believed that animals housed reincarnated human, therefore rational, souls.[4] The Cynics claimed that animals were superior beings. It was not until the fourth century B.C.E.

that Aristotle denied animals' rationality and, in doing so, provoked "a crisis both for the philosophy of mind and for theories of morality," the implications of which are still contested today (Sorabji 1993, 7).[5] At the risk of oversimplifying a highly nuanced philosophical system, Aristotle granted animals a rich sense of perception but denied them reason, thought, intellect, and belief. In *Politics*, he divided the world into those who could plan their own lives and those who could not. He ranked living creatures on a "ladder of life" according to their rational abilities, with humans at the top and inanimate things at the bottom. Following Plato, he also differentiated humans into higher and lower groups, with those possessing lesser intellects (than Greeks) destined for slavery. Animals allegedly had even less intellectual ability than slaves did, and, because nature does nothing in vain, they had been "naturally" designed to serve the more perfected humans. The Stoics further sharpened the human–animal distinctions in the third century B.C.E., when they denied legal protection to animals because they could not give or withhold assent. Stoic theory percolated into Christian doctrine during the fourth century C.E. through the writings of Augustine. In the *City of God* (1.20; written by 413), Augustine insisted that the commandment "Thou shalt not kill" did not extend to animals. "Because they are not associated in a community with us by reason," he wrote, "[their] life and death is subordinated to our use." Moreover, Augustine interpreted one of the acts attributed to Jesus as support for the Stoic doctrine. The New Testament contains an account of Jesus casting demons out of a man and into a nearby herd of swine, who consequently drove themselves over a cliff and into the sea (see Matthew 8:28–32; Mark 5:1–17). With this act, Augustine argued, Christ himself validated the Stoic view that animals did not belong to the legally protected community. Lawrence argues that the importance of distinguishing humans from animals in the early church had to do with distinguishing pagans from Christians. Because animals in the ancient world could easily change into humans, and vice versa (Ovid's *Metamorphosis* is a good example), early church fathers "rejected such species ambiguity and firmly established the doctrine of qualitative differences between people and animals" (Lawrence 1995, 76).

In the thirteenth century C.E., Thomas Aquinas solidified rationalist, anti-animal, Catholic dogma.[6] Thoroughly trained in Aristotelian thought, Aquinas maintained that the only part of the soul that survived after death was the part that reasoned. Lacking this capacity, the souls of animals died along with their bodies. With this formulation, Aquinas relieved Christians of having to treat animals with kindness, because they would not meet the creatures they had exploited in the afterlife. Granted, he advised humans against outright cruelty, but not because of its inherent evil. Rather, its harm lay in its potential to lead to the more serious offense of cruelty to other humans. This perspective, known as the indirect duty view, has currency to this day.[7]

Aquinas symbolizes a "breathtakingly anthropocentric spirit" (Thomas 1983) that was enforced through violence, torture, and execution.[8] On the European continent, this came in the form of the Inquisition, the agency commissioned in 1231 by Pope Gregory IX to investigate heresy. Although this was the Inquisition's official purpose, "its main goal appeared to be the eradication of anyone or anything that contradicted the biased, hierarchical, Aristotelian/Thomist view of man's place in nature" (Serpell 1986, 155). Consequently, the Inquisitors authorized themselves to decide which relationships with animals were appropriate and which were heretical. Although animals figured into countless aspects of daily life, companionship with them meant a perverting of the anthropocentric hierarchy, and accusations of bestiality and witchcraft were common—and uncontestable. The Inquisitors also oppressed and, when possible, obliterated the numerous, popular nature-worshipping cults, which the church had previously ignored. Moreover, a number of saints whose images had been depicted with animals during the early Christian era were revised during the Inquisition. Saint Christopher and Saint Bernard are two of the most noted. More striking is the case of a dog who was a saint—at least, until the Inquisition. In the region around Lyons, France, a cult had formed around Guinefort, the greyhound saint (see Thomas 1983; McDonogh 1999). The legend of Guinefort maintains that the dog's master came home one day to find his infant child missing from his crib and both the crib and the dog covered in blood. Assuming that the dog had killed the child, the outraged master killed Guinefort. Afterward, he learned the truth. He saw the torn remains of an enormous snake that had menaced the child, who, thanks to loyal Guinefort, was found sleeping safely nearby. The remorseful master placed Guinefort's remains in a well and planted a grove of trees in the dog's honor. The grove became a sacred site, attracting people from miles around who brought their children for the healings that the dog, now Saint Guinefort, performed—until church officials had his remains disinterred and burned, along with the sacred grove. Guinefort's legend is one that survives; surely, countless other examples of the harmless worship of nature and animals vanished from the record as the Inquisitors sought to destroy all threats to the fragile distinction between humans and animals.

The effort exerted to maintain the human–animal boundary itself constitutes evidence of the boundary's artificial—and political—quality. If the line between humans and animals were indeed "natural," it would not have required violent enforcement. However, the violence shows the lengths to which groups that want to increase their power—in this case, the church—would go. I am not arguing that distinguishing humans from animals was wrong in principle. My point is that distinguishing between the two was less about delineating equal species than about giving humans special status and justifying dominion over nature. In addition, the debate over the human–animal boundary is a lens into the debate over acceptable human conduct. Aristotle, Augustine, Aquinas, and others sought to identify what mattered in life (see Sorabji 1993, 218). Their arguments responded to contemporaneous social issues, such as the justification of slavery and the basis for law. In creating positions on these issues, they made things manageable by reducing the scope of their considerations and placing animals beyond the limits of concern. This is the sense in which the human–animal boundary is a social construction, which does not mean that it is not real. Rather, it means that it served as a conduit for the concerns of people of the time, from justifying dominion to determining the reach of the law. I have mentioned that other views of animals were in circulation at the time, and one of these could just as easily have become the dominant view. For example, Theophrastus, one of Aristotle's students, insisted that animals deserved justice and that it was wrong to cause their suffering, even by eating meat. However, the rationalist view became "reality" because it legitimized the existing social order.

NEW CATEGORIES: CLASS, STATUS, AND PETS

Social order is always precarious. Because meaning is neither natural nor objective, new patterns of social interaction and new expectations give rise to new meanings. The socially constructed character of the human–animal boundary becomes more evident, considering that certain groups of people could defy theological dogma and live closely with animals. Granted, they needed the resources to support animals who had no economic function, but, more important, they needed the social status to protect them from the consequences of doing so. The first to have these qualifications were the ecclesiastical elite and the nobility. As these groups allowed certain animals into their midst, the function of the human–animal boundary changed in several ways. It became not just about setting humans above animals, but also about setting certain people above other people. In important ways, then, the fragile boundary between humans and animals began to define social classes. Comparisons with animals were symbolic ways to say that another person was subhuman. For example, in a sixteenth-century text on civility, Erasmus distinguishes good manners from bad based on those that make people into animals, for only horses smack their lips while eating and only dogs chew on bones and show their teeth. Curbing the "animal" or the "brute" within the person was essential in an era fraught with anxiety about maintaining the human–animal boundary. In addition, the boundary flexed enough to define new categories of animals, as the dog gained an almost human status, the cat became diabolical, and the "pet" emerged as a ubiquitous, albeit controversial, symbol of status.

Officially, religious orders prohibited their members from keeping animals except the occasional cat to keep away vermin. Records of church councils show that dogs were widely forbidden in monasteries as early as the sixth century (see Menache 2000). There were various reasons for the prohibition. First, early Christian doctrine had absorbed the disdain of dogs prevalent in the Middle East, where dogs were considered unclean (and remain so today; see Menache 1997). The Bible is full of references to dogs as "filthy." In the New Testament's Book of Revelation, this status excludes dogs from the resurrection and eternal life in the New Jerusalem.[9] Such creatures could hardly associate with nuns and priests. Second, church authorities had determined that hunting was a "carnal diversion," inappropriate for members of the clergy (see Thomas 1983; Menache 2000). Because the existence of animals was justified by their usefulness to humans, and because the clergy did not hunt, they had no need for dogs. Third, church authorities alleged that feeding animals diverted alms that should have gone to the poor, and fourth, that people might be afraid to approach the residence of a cleric who had dogs. Despite the strict prohibitions, however, evidence reveals that monks, nuns, and the lay people who chose monastic life nevertheless kept phenomenal numbers of animals of all kinds as companions. In addition, images of dogs and other animals appear frequently in illuminated manuscripts created by monks. Some monasteries may even have produced their own breeds of dogs (see Menache 2000). The devotion with which religious authorities pursued the issue of animals in holy orders shows the difficulty they had enforcing rules about human supremacy among their own ranks. Thus, even in the theological arena, which most energetically policed the human–animal divide, the boundary was contested. The intrinsic rewards of relationships with animals were apparently worth the risk.

The nobility also escaped the stigma of association with animals, particularly dogs. As hunting evolved into an elite (male) pastime, rather than a necessity for survival, it legitimized the dog's entry into human circles (see Serpell 1988a; Menache 2000).[10] By the end of the thirteenth century, when hunting had lost its subsistence role for the elite, distinct, class-specific patterns of behavior emerged. Among the nobility, hunting became a sport, which is to say it became a pursuit embedded with the virtues of courage and bravery allegedly found only in the noble constitution. Because dogs were associated with successful hunting, and because successful hunting was a necessary status symbol, dogs became ubiquitous in the upper ranks of society. Sophia Menache writes of a "socio-economic, cultural equation of nobility = hunting = dogs" (2000, 55; see also Cartmill 1997). This equation elevated dogs out of the animal kingdom and into a respectable place in the nobility purely through their ability to improve the chances of a successful hunt. Illuminated manuscripts dating from this era depict how, "after the dog had become a condition sine-qua-non for this success, it was dissociated from other animals and invested with a unique place in human society" (Menache 2000, 56).[11] Once this occurred, the privileged status seems to have dispersed from hunting dogs to dogs in general, even those who did not hunt. For example, royal portraiture reveals that small dogs became ubiquitous at European courts around the fifteenth century. The first portrait to include a small pet dog, decidedly not a hunting dog, is *The Arnolfini Marriage*, painted by Jan van Eyck in 1434.[12] The instrumental view of animals that was characteristic of the time meant that dogs existed to serve people. Thus, new jobs emerged for new breeds: Some caught rats, while others, in the official role of *chien-goûteur,* tasted the food to prevent royals from being poisoned. Some had the job of alerting their royal owners of the presence of intruders in the bedchamber. Many, however, were solely companions. Lap dogs, obviously useless for hunting, became fashionable first among women. Their uselessness and popularity with women combined to make them easy targets for criticism. Whereas large breeds, such as the greyhound and mastiff, were approved symbols of virility, the smaller dogs connoted femininity and impotence. Although the breeds depicted in artwork are impossible to identify precisely today, many appear to be toy spaniels. Even these smaller dogs, like their hunting kin, had functional roles, such as keeping fleas away and providing warmth and consolation—hence, the name given to them: "comforters." Some critics, however, had an unshakable image of a "real" (read "masculine") dog, which the "comforter" did not fit. For example, John Caius, physician to the court of Henry VIII and author of the first English book on dog breeds, disliked small breeds because they were "chiefly for the amusement and pleasure of women"; his contemporary William Harrison called them "instruments of follie" that caused women to "[trifle] away the treasure of time, to withdraw their minds from more commendable exercises," and, worst of all, "to content their corrupt concupiscences with vaine disport" (see Ritvo 1988; Serpell 1988a; McDonogh 1999). Upper-class women apparently ignored such criticism, because the "useless" breeds of dogs continued to gain popularity until "no well-to-do woman was complete" without one (Thomas 1983, 108). The pet had arrived.

Those without money and, more important, without rank faced a different situation. Prejudice rather than poverty restricted pet keeping to the upper classes. Among the poor, hunting remained a necessity, even as new game laws about when, where, and how it could take place made hunting increasingly difficult (see Cartmill 1997; Menache 2000). As the lower classes were edged out of hunting legally, statutes simultaneously made it unlawful for them to own particular kinds of dogs—not surprisingly, those typically used for hunting (see Thomas 1983; Derr 1997; Menache 2000). For example, twelfth-century English laws made it illegal for anyone except the nobility to own mastiffs, spaniels, and greyhounds. Others who kept dogs most often had multipurpose mongrels known as "curs." The only way around the laws against particular breeds was shockingly inhumane. Non-elites could own mastiff-type dogs for protection but only if the dogs were "expeditated" or "lamed" (see Derr 1997, 54–55). The dogs' front paws were placed on a thick block of wood and, using a chisel and mallet, the middle claws were chopped off at the flesh. This amputation prevented them from chasing the forest owner's game. Granted, some people could not afford to own animals who had no economic function, but even those who could do so faced risks. The potential accusation of witchcraft and bestiality loomed constantly.[13] The accused—usually women, usually poor and elderly—were often suspected solely because they "possessed and displayed affection for one or more animal companions" (Serpell 1986, 57; see also McDonogh 1999). Their animals went to death with them, because they were most likely "familiars"—diabolical companions who carried out evil deeds in exchange for food and shelter.

This was particularly the case for cats. Whereas the dog, thanks to hunting, had gained a status on the borderlands above animals and closer to humans, the status of cats went in the other direction. Once so valued that the Egyptians prohibited their export, cats became a symbol of everything demonic. Their downfall began with the spread of Christianity and the simultaneous extermination of pagan forms of worship. Widespread persecution of cats began with the Inquisition, when killing them—especially black ones—became a holy mandate. Although we can never fully know what caused the hatred of cats, surely their enigmatic natures and their quiet, slinking movements—the very things that endeared them to the Egyptians—threatened the anthropocentric view. As a result, the church condoned the torture and execution of cats on saints' days and other holy days. For example, the French regularly burned cats to death at the festival of Saint-Jean, which marked the end of the planting season (see Kete 1994; McDonogh 1999). Cats also went to their deaths in rites of purification and protection. A Lenten ceremony in Metz involved placing thirteen cats in an iron cage and burning them to death (see Kete 1994, 119). The event occurred annually until 1777, with a similar occasion recorded in Lorraine as late as 1905. Robert Darnton (1985) and Norbert Elias (1994) offer other examples of the burning, torture, and boiling of cats. Well into the seventeenth and eighteenth centuries, cats remained easy targets for torture. The English often stuffed effigies with cats who screamed as they burned, providing sound effects that cost them their lives (see Thomas 1983; McDonogh 1999).[14]

In sum, medieval and early modern attitudes toward dogs and cats existed within the context of a nascent class system and a powerful church that was openly antagonistic to animals. Christianity could not tolerate close contact between humans, who were made in God's image, and animals, who were made to serve the needs of humans. However, dogs occupied a unique category because they helped humans in the killing of other animals. When the emerging elite began to gain power formerly held only by the church, aristocratic men and women could associate with dogs without fearing the consequences suffered by the less affluent. The elite could justify associating with dogs because they were useful for hunting. However, the fragility of the social-stratification system becomes clear in the criticism launched at small, "useless" dogs and in the restrictions placed on dog ownership among non-elites. It is also evident in the demonization and persecution of cats, who served to set apart those who threatened the human–animal boundary in numerous ways.

The unique status given to dogs constitutes a watershed in human–animal relations. However, medieval and early modern relationships between humans and dogs probably did not involve close emotional ties. The pervasiveness of the doctrines of human supremacy and dominionism and the attendant definition of animals in instrumental terms, combined with the power of the Catholic church to enforce these views, leave little reason to suspect that people could feel affection for animals in the sense that we understand it today. Although people clearly admired and appreciated dogs, the divorce between the human and animal kingdoms was severe enough to preclude any attempts to understand what dogs thought or felt, or to imagine that they could think or feel at all. Those sorts of relationship would become possible only when science reduced the distance between humans and animals, and even then—and even now—the continuity between us and other species would remain highly contested.

SCIENCE CHALLENGES THE ANTHROPOCENTRIC ILLUSION

As the Western world advanced into the modern era, the debate about the differences between humans and animals took on scientific, rather than solely theological, terms. Granted, theological arguments remained in the picture, and would do so until—and despite—Darwin. However, the debate revolved primarily around the issue of rational abilities, albeit cloaked in questions of a creator's design. For example, in the seventeenth century, the status of animals was informed by René Descartes, who portrayed them as automatons, so thoroughly "other" than human that they lacked consciousness of their own pain. More accurately, Descartes argued that animals' pain was insignificant because they lacked self-conscious awareness of feelings. They merely responded to physical stimulus. Although humans were considered automata, as well, they differed from animals in having additional abilities, including consciousness and souls, which we could manifest through speech. Put more clearly, "Descartes famously thought that nothing short of conversational ability in a human language could support the attribution of consciousness to animals" (Allen and Bekoff 1997, 144).

As Keith Thomas (1983, 34) explains, "The most powerful argument for the Cartesian position was that it was the best possible rationalization for the way man actually treated animals." The Cartesian view justified vivisection without the use of anesthetics and other horrific though commonplace cruelty. Of course, one need not have known the details of Cartesian thought to understand that a gulf existed between humans and animals. Everyday life was full of reminders of animals' inferiority, for exploiting their labor required extensive justification practices. Undesirable behaviors were labeled "brutish" and "bestial" and undesirable people were "animals," ruled by desires and urges. Although we still hear animal insults today, they were surely far more potent in an era devoted to distinguishing between human and not human. For most people, animals symbolized the harsh, depraved, unclean natural world, and bringing one into one's home on an equal basis would have required an attitude that few ordinary citizens had at the time (see Ritvo 1987, 1988). The paucity of written evidence of pets in the American colonies itself speaks volumes on attitudes toward animals. What little evidence does appear takes the form of satire, disapproving and suspicious of those who felt affection toward animals who had no economic function. Among the American and British elite, however, lap dogs and hunting dogs were popular, judging by their frequent appearance, along with horses, in formal portraits dating from the seventeenth century. It is not clear, however, whether these animals were pets in any sense of today's usage of the word. Perhaps they functioned as semiotic devices in portraiture, equivalent to other pieces of equipment that portrayed the masters' social status. In any case, the upper classes continued to escape the stigma of association with animals, especially dogs.

By the late seventeenth century, the anthropocentric tradition had begun to face challenges on several fronts. For one thing, the flawed logic of Cartesian view brought on its own demise. As Serpell (1986, 161) explains:

> The Cartesian vivisectors had sowed the seeds of their own destruction. All the evidence that they had accumulated on the internal anatomy and physiology of animals merely served to emphasize their similarity to humans. And if the underlying mechanisms and responses were the same, then it was highly probably that animals and humans both experienced similar sensations of pain and discomfort.

The Cartesian view had wanted it both ways. It portrayed animals as machines, not only different from humans but also inferior to them. Yet despite these differences and inferiorities, it justified using animals as surrogates for humans in medical and scientific experiments. Even in its time, Cartesian anthropocentrism had its opponents. Voltaire ([1962], 113), for example, wrote: "You discover in him all the same organs of feeling that are in yourself. Answer me, machinist, has nature arranged all the springs of feeling in this animal in order that he should not feel?" Descartes had numerous other opponents, including John Locke and the theologian Henry More, who called the Cartesian view "murderous." The crucial blow came in the form of new systems of biological classification. Previously, animals (and plants) had been categorized according to their utility to humans (that is, edible–inedible, tame–wild, useful–useless). Then the biologist John Ray, along with Linnaeus, developed taxonomical systems based on similarities in physical structure. This meant seeing all of nature with more detachment, regardless of its usefulness to humans. Additional challenges to anthropocentrism came from geology, which offered the knowledge not only that the Earth was much older than had been assumed, but that it had existed long before the appearance of humans. Likewise, astronomy proclaimed that the universe was perhaps infinite; the microscope revealed worlds in a drop of water; and scientists in several disciplines began to discuss versions of evolutionary theory in the form of a "Great Chain of Being." In short, the era witnessed a "revolution in perception" (Thomas 1983, 70) from which the anthropocentric illusion would never fully recover.

News of the anatomical and biological similarities between human and non-human animals was first available mainly to the elite (see Tester 1992). The cultural effect of this new information was a greater respect for life, as shown in the philosophical and theological discourses of the Enlightenment. For if animals and humans were similarly constituted, then animals could feel. And if they suffered pain just as we did, then human conduct toward our "fellow creatures" was barbaric. In the mid-eighteenth century, Jean-Jacques Rousseau wrote about animals and humans as equally sentient beings, both deserving of respectful treatment. In 1781, Jeremy Bentham (1988 [1781]) penned his famous response to the subject of cruelty to animals:

> A full-grown horse or dog is beyond comparison a more rational, as well as a more conversable animal, than an infant of a day, or a week, or even a month, old. But suppose the case were otherwise, what would it avail? The question is not, Can they *reason?* nor Can they *talk?* but, Can they *suffer?*[15]

In stark contrast to previous eras, religious authorities supported the new sensibilities. The issue of animals having souls or reason became inconsequential, for the capacity to feel was argument enough against mistreatment. It became "unnatural" to stand idly by while other creatures suffered, and the first legislation against cruelty began to appear. In 1822, Britain passed the Ill Treatment of Cattle Act, known also as Martin's Act, which was the first legal measure aimed at animal protection. The British SPCA was founded two years later (it would gain permission to use the "Royal" prefix in 1840). Several German states meanwhile enacted anti-cruelty laws.[16]

The greatest setback to anthropocentrism came with the publication of Charles Darwin's *On the Origin of Species* (1859). Whereas others before him had postulated that life had evolved, no one had yet established the mechanism by which the process of evolution occurred. The theory of natural selection offered such a mechanism and in doing so challenged religion and anthropocentrism. For unlike Christian and Aristotelian theories, which argued that a supreme creator or God had designed the various species to fulfill certain (human) purposes, natural selection had no designer and no purpose. Darwin made the idea of a creative being or power irrelevant and knocked humans from their pedestal of superiority. Later, in *The Descent of Man* (1871 [1936], 448), Darwin emphasized kinship among species, stating that "there is no fundamental difference between man and the higher mammals in their mental faculties." Extending this claim's moral implications, he wrote that humans would reach the highest point of evolution when our concern "extended to all sentient beings."

The idea of a kinship among species spawned two interpretations, each with different implications for the treatment of animals (see Franklin 1999).[17] One view held that if humans were part of nature, then activities such as hunting and sport fishing were justified because they united humans and animals in the competitive, even violent, ritual of life and death. In this view, an activity such as big-game hunting could invigorate mind, body, and soul made flabby by modern life.[18] The wholesale slaughter of African and North American wildlife in the second half of the nineteenth century suggests that a considerable amount of "invigoration" took place. The other view maintained that if humans were indeed part of nature, then they ought not to exploit it but must instead preserve it. The latter view manifested itself in a wave of efforts to protect animals.[19]

At the time, "the most appalling cruelties were often associated with the most routine economic activities" (Ritvo 1987, 137). The epitome of nineteenth-century treatment of animals was the beaten and bedraggled horse. A common means of getting horses to pull impossibly heavy loads was to light a fire under them. They often dropped dead on the streets from exhaustion, and not until very late did they even have watering stations at which to drink. One of the most powerful artifacts that was both cause and consequence of the changing attitudes was *Black Beauty*. Published by Anna Sewall in 1877, the book illuminated the plight of horses on both sides of the Atlantic. In addition, the increasing number of organizations and of legislation against cruelty provide documentary evidence of changing attitudes. London banned the use of dogs to pull carts in 1840. The Société protectrice des animaux was founded in Paris in 1845. In the United States, the Civil War slowed animal-protection efforts, but immediately after the war, in 1866, Henry Bergh founded the ASPCA. A law prohibiting cruelty to animals passed the next week, with the ASPCA having authority to enforce it. Groups that opposed vivisection formed in Britain, France, and the United States during the 1870s. The American Humane Association was established in 1877.[20] In the United States, the first humane shelters—as opposed to pounds, which held animals for a time and then killed them or sold them to vivisectors—also appeared in the 1870s.[21] Before the Civil War, animals picked up as strays were held for a time (which depended on the municipality) and then killed, usually by electrocution, or sold in quantity for vivisection.[22] In the 1860s, two Philadelphia women, Elizabeth Morris and Annie Waln, began picking up and sheltering strays from around the city. They found homes for some and chloroformed those for whom they could find no homes. In 1874, Morris and Waln started the first Animal Rescue League. In 1888, their organization was independently incorporated as the Morris Refuge Association for Homeless and Suffering Animals. It became a model for many humane shelters and still operates on Philadelphia's Lombard Street. At about the same time, Caroline White, another Philadelphian, grew outraged at the conditions of the city pound and its role in supplying animals for vivisection.[23] White had spearheaded the formation of the Pennsylvania Society for the Prevention of Cruelty to Animals (PSPCA), but as a woman she could not hold an office (although her husband was on the board). In its second year, the PSPCA proposed the creation of a "Women's Branch," and White became president, a position she held until her death in 1916.[24] Among the many humane accomplishments of the Women's Branch was that it initiated the first contract with a city government to shelter strays humanely. The city contributed $2,500 toward the costs. This constitutes "the first attempt on the part of any society in the United States to handle the problem of caring for surplus or unwanted small animals, and as far as it can be ascertained the first appropriation ever made by a municipality for humane work" (Coleman 1924, 181; see also Brestrup 1997). This system, whereby an independent humane society contracts with a city to do "animal control," is still in practice.

The humane movement represents a shift in the designated overseers of the human–animal boundary, as well as a subtle change in the terms of the associated debate. When the terms were religious, the boundary was the church's domain. When science began to secularize the terms of the debate, the middle class staked a claim to that territory. The debate was less about distinguishing animals from humans than about distinguishing the types of humans had the ability to treat animals properly. This, too, is an artifact of evolutionary theories. Darwin had blurred the boundary between humans and animals, but various popular forms of "social Darwinism" supported beliefs that some groups of people were more highly evolved than were others. In this case, anti-cruelty efforts reveal middle-class attitudes about the inhumanity of the lower classes. For instance, the founders of the Royal Society for the Prevention of Cruelty to Animals framed their efforts on behalf of animals as a means of disciplining working-class behavior (see Ritvo 1987). Although the crusaders garnered extensive popular support against cruelty, they faced difficulty mobilizing legislative support—and they desperately needed elite backing to do so. Framing the issue as social engineering was the key. For during this time, droves of the rural working class were moving to the cities. Urban working-class occupations frequently exploited animals' labor, often through excessive cruelty. Furthermore, much working-class "entertainment" of the time consisted of violent contests between animals, such as the baiting of bulls and other animals, dogfighting, and cockfighting.[25] The urban middle class grew increasingly fearful of encouraging the potential violence and disorder that would emerge from groups already thought to be violent and disorderly. This fear was precisely the card that the cruelty reformers played. By taking the indirect duty view that the mistreatment of animals caused cruelty to humans and by pointing out the already brutish nature of the lower classes, middle-class reformers gave their cause legislative appeal.[26]

I raise this point about the targeting of the working class not to cast doubt on the authenticity of the anti-cruelty crusaders' intentions. Rather, I want to highlight how animal-protection efforts were a vehicle for sustaining power relations between dominant and subordinate groups.[27] The struggle would continue at another level as the pleasures of animal company opened to people of all classes.

PET KEEPING DEMOCRATIZED

During the second half of the nineteenth century, numerous factors coalesced to make pet keeping a widespread practice. For one, Darwin—and science, in general—had helped make animals less threatening and more interesting. Simultaneously, as dominion over nature became more complete, people could safely allow select forms of nature into their homes.[28] Rates of pet keeping skyrocketed in Europe and the United States, and numerous cultural artifacts document the institutionalization of the pet during the late nineteenth and early twentieth centuries. One of these is the "pet store." In her study of pets in Paris, Kathleen Kete found that until about 1880, people who wanted to obtain dogs could do so mainly by buying them at the Sunday horse market or through private breeders. Gradually, vendors began to place advertisements in newspapers. By 1910, Kete found, numerous print advertisements were appearing for shops that sold purebred puppies, along with dog collars and accessories. The first animal cemeteries in the United States, founded in 1898 and 1906 (see Coleman 1924), attest that pets had become established parts of middle-class society, as does the commercial production of dog food. The British company Spratts Patent was the first to combine cereal, vegetables, and a small amount of beef into a biscuit. For toy breeds, Spratts produced a meatless version. The pellet or nugget form of dog food known today did not appear until several decades later. The American pet-food industry started with Ralston Purina, founded in 1894 as a horse- and mule-feed business called the Robinson-Danforth Commission.[29] In the 1920s, Purina began selling the first commercial dog food, which was available through rural feed outlets.[30] Up to then, dogs ate table scraps and anything that humans did not eat. A 1917 pet manual suggests feeding dogs "a certain percentage of meat, preferably cooked. . . . To this may be added various vegetables, bread, cooked cereals and milk, in fact almost anything edible, provided it is clean and not too greasy" (Crandall 1917, 6). The recommendation for cats is similar. Their diet should consist of "meat, either raw or cooked. It is not necessary that this be of the finest quality, but neither should it be of the vile sort sold by butchers as 'cat meat'" (Crandall 1917, 14). One necessity for life with cats—namely, litter—was not developed until after World War II (see Maggitti 1996).[31] Until then, people filled their cats' litter boxes with sand or ashes.

As I mentioned earlier, the establishment of the first humane shelters is another artifact of the late nineteenth century. Their appearance represents acceptance of the attitude that dogs and cats belonged in homes, with human families, and that roaming animals had become a social problem, which reveals much about the ideological underpinnings of urban life. Shelters also contributed to the rise in pet keeping by enabling people who were not of the upper class to obtain companion animals (see Coleman 1924). Shelters became places where people who could not afford pet stores or breeders or preferred to rescue secondhand animals could go.

Despite the increasing democratization of pet ownership, animals still conveyed different meanings depending on the owner's status. Purebred dogs became all the rage among the middle and upper classes. The breeding of dogs itself became fashionable, and most of the breeds known today date from the Victorian era. Lynda Birke (1994, 36) points out that breeding "generate[d] animal mimics of human (Victorian) society, complete with elite aristocrats—the well-pedigreed winners in the show ring—and useless miserable curs." The pets of the lower classes took the blame for all manner of physical and moral dysfunction. As Ritvo (1987, 179) details, the dogs of the English poor were accused of being more vulnerable to rabies than "well-bred dogs." To combat rabies and control the dog population, dog licensing and taxes became standard practice in many American and European cities. The regulations usually applied to dogs older than six months of age and limited ownership to those for whom dogs were economically necessary or those who could afford the licensing fees for "luxury" dogs (see Thomas 1983; Ritvo 1987; Kete 1994). To avoid the fees, many people simply let their dogs loose in the streets once they passed the endearing puppy stage. Consequently, in most towns and cities, packs of dogs ran free, creating numerous potential and actual problems.[32] Granted, many middle- and upper-class owners neglected to license their dogs, too, but the "useless miserable curs" of the lower classes made easy scapegoats. Ritvo offers numerous accounts from London newspapers accusing strays of behaving as badly as the people who had released them allegedly did.

Further evidence of people of lower incomes' keeping pets, as well as middle-class efforts to curb them from doing so, appears in the prohibition of pets in rental apartments. Late Victorian public housing forbade tenants from keeping dogs, implying that the poor were incapable of caring responsibly for them (see Jones 1971, 186). However, people of all classes managed to enjoy the pleasures of canine company. When Libby Hall (2000) documented a late–nineteenth-century trend to have dogs photographed, she traced many photos to studios in low-income London neighborhoods, suggesting not only that companion dogs were present among the poor but also that the relationships sometimes merited costly commemoration.

The most striking example of the incorporation of pets is the case of the cat. Branded as the icon of dangerous, feminine sexuality, the cat's image was "rehabilitated" in the mid- to late nineteenth century (Kete 1994, 117, 127). Because cats had retained much of their wildness, they had long resisted categorization as "pets" in the strictest sense of the word (see Griffiths et al. 2000). In contrast to dogs, those symbols of fidelity and affection, cats sent another message. In nineteenth-century Parisian culture, for example, cats were considered neither working class nor bourgeois; they were associated with bohemians, especially intellectuals and artists (see Kete 1994). Cats were the *anti-pet,* with their languor and comfort-seeking embodying the alternative to urban, bourgeois life. By the turn of the twentieth century, however, cats had become widely popular domestic companions. Several factors went into this rehabilitation. First, the cat's qualities gradually "lost the power to reassure or disturb" (Kete 1994, 134). The independence, promiscuity, aloofness, disloyalty, indolence, and other characteristics long attributed to and feared in cats were recognized for the inaccurate projections they were. Moreover, friends of cats championed feline qualities as enjoyable in themselves. Second, and related to this, modernization encouraged the incorporation of safe elements of the "exotic" into everyday life. Although an "ordinary" cat would accomplish some of this, one of the breeds popularized in the era would go much further: Siamese, Abyssinian, and Persian—even the names suggest the unusual and the mysterious. Third, in an era that was becoming newly aware of germs, the cleanliness of cats set them above dogs. Whereas the cat's fastidiousness had been a target of mockery and criticism, it now made them seem preferable to dogs, who were equally happy whether dirty or clean and would eat garbage just as enthusiastically as they would eat roast beef.

The inclusion of cats as pets was in many ways a last hurdle in the race to dominion over nature. Whereas people had once had to justify keeping animals who had no economic function, the "pet" represents a caste of such animals, whose sole purpose is to be human companions. Both dogs and cats became the objects of highly selective breeding, the dog more successfully than the cat, and both were exhibited at shows that drew crowds of "fanciers." In addition, this breeding aimed to reduce traces of the "pet's" animal nature. Pets could look wild and exotic, but they had to behave like proper members of the family. The "pet," as Shepard (1978) puts it, was a "minimal animal"—never aggressive or sexual, always cheerful, quiet, playful, and happy.

3
From Pets to Companion Animals

When an animal is thought of as a piece of clockwork machinery, then some of its most interesting attributes are almost certainly overlooked.

—Patrick Bateson and Dennis C. Turner (1988, 200)

If we could go back in time to survey Western people's views of animals across the centuries, I am certain of one thing: Most would not describe animals as thinking, feeling partners in social interaction. Some would surely express great affection for animals, but few would describe animals as having selves akin to the human experience of selfhood. By the late 1990s, however, that would have changed enough for scholars to begin taking notice. In *Understanding Dogs*, Sanders (1999, 3) found that a majority of people who lived and worked with dogs defined them "as thoughtful, reciprocating, emotional beings with uniquely individual tastes and personalities." The cat guardians studied by Alger and Alger (1997, 2003) offered similar reports.

In the previous chapter, I argued that the practice of pet keeping represented a particular understanding of animals that became widely possible only under certain social, cultural, and economic conditions. The next development, too, would depend on a similar series of changes. During the 1990s, the term "companion animal" became a popular, albeit contested, alternative to the word "pet." Although sometimes used synonymously, companion animal has a connotation that is quite different from "pet," which evokes images of Tricky Woo. "Companion animal" has a political context, connoting an increased effort on the part of humans to accept some animals as animals rather than as workers, decoration, or entertainment (see Franklin 1999, 86). "Companion animals" remain "other" than human but in a sense worth honoring, rather than one of inferiority. I offer several disparate examples of this new sensibility:

- New knowledge about animals' behavior, emotions, and cognition;
- Increased awareness of the problem of pet overpopulation, resulting in affordable spay and neuter programs;
- Increased visibility of homeless animals to encourage adopting from shelters instead of purchasing from "puppy mills";
- Technological advances such as the implanted microchip, which reunites stray animals with their guardians;
- Veterinarians who practice holistic and alternative medicine (see http://www.ahvma.org);
- Veterinarians who operate what are called "bond-centered" practices;
- New activities for animals (especially dogs) and guardians to engage in together, such as agility and flyball;[1]
- Increased awareness of the harm of surgical amputation and mutilation through declawing, de-barking, tail docking, and ear cropping.

Perhaps the clearest illustration of the change comes from the realm of dog training, particularly in the trend toward nonviolent, compassionate methods that has emerged over the past fifteen years. "Traditional" training involves "leash corrections," meaning jerking or snapping the dog's leash, which is usually connected to a collar that chokes, pinches, or even shocks. In many instances, owners whose dogs do not understand the first few times end up lifting them off the ground—hanging them, to be precise—until they listen. This philosophy became standard practice in the 1960s and early 1970s, largely through the work of the trainer William Koehler. His methods include the "scruff shake" (picking the dog up by the skin on the back of the neck and shaking him or her), the "alpha roll" (forcing a dog onto his or her back in a submissive position), and "helicoptering" (swinging the dog by choke collar and leash—for those really stubborn dogs).

In traditional dog training, teaching a dog to sit would proceed along these lines. A person sternly tells a dog, often a puppy, often outfitted with a choke collar and leash, to "Sit." Having no command of English, the dog has no idea what the word means and simply stands there looking around at the many distractions the world offers. The person thinks the dog will not listen. He or she jerks the leash, sending a jolt to the dog's windpipe to get his attention and says, "Sit!" a few more times. The dog still has no idea what the word means and looks around again, only to receive another jerk, this time followed by an upward yank on the leash that knocks him off his feet, and then another stern "Sit!" from the human. By this time, the confused dog has stopped enjoying the training session. The person, not enjoying it much either, pushes the dog's rump to the floor, jerks his head upward and, nearly shouting, says again, "I said, 'Sit!' What's wrong with you, you stupid dog?" What the dog has learned from this is to fear the person. Moreover, when his rump hits the ground, he is spoken to harshly, so he is understandably reluctant to offer that behavior on his own. It appears to the person as though the dog will not listen. The person becomes angry. The dog remains confused. After several such sessions, the human gives up and ties the dog to the tree in the backyard or locks him in the garage. The dog, bored and lonely, takes up recreational barking. The neighbors complain. Because the person and the dog essentially have no relationship, it is not so difficult to take the dog to a shelter.

Granted, many dogs do learn to sit, and more, through traditional training. Moreover, not all of them end up like the one in my example. I have overstated the case to make two points: first, that traditional training expects dogs to learn in human terms, by responding to language that dogs do not understand, and second, that traditional training is solely about getting dogs to do things, such as sit. In contrast, here is another scenario. This time, the person recognizes that the dog has no knowledge of English. The person has a supply of rewards for the dog, perhaps tiny pieces of cheese or meat or a toy for those who are not motivated by food. Holding some of the treats in one hand so the dog can see and sniff them, the person first gets the dog's attention. He or she then raises the treats over the dog's head. The dog looks up, and the person gradually moves her hand back. The dog's head follows, and as the handful of treats moves farther back, the dog's natural movements lure him into a sitting position. As the dog's rump hits the ground, the person says "Yes!" or otherwise marks the good behavior, often using a clicker, and gives the dog the treat. The word "sit" has not yet entered the picture. The dog and person practice a few more times. Together, they are "shaping" behavior.[2] The treats flow steadily. Before long, the dog makes the connection and sits on his own, and the person can then introduce the word because the dog can now associate it with a behavior. The person can slowly discontinue the treats over time, ensuring that the dog will offer the behavior under any circumstances. The dog learns without force, through language he can understand. The dog and the person have created a partnership in which their minds and bodies work together. They watch, listen to, and generally "read" each other. They enjoy each other's company. The training is not about getting the dog to obey—or, at least, it is only partly so. It has more to do with creating a relationship.

These examples illustrate two dramatically different ways of thinking. One is "obedience training," and the other is a combination of positive reinforcement and behavior modification. The former approach legitimizes physical violence and abuse disguised as discipline (see Derr 1997, 327). The latter approach draws on principles of behavior modification and the philosophy of nonviolence. The former approach represents the view that animals are simply bundles of instincts. The latter maintains that they possess a level of consciousness that makes them similar to humans in many ways. This idea is crucial in accounting for the transition from "pets" to "companion animals." As we saw in the emergence of pet keeping, numerous factors were at work in the transition.

EXPLAINING NEW ATTITUDES

To explain the change, I turn first to the work of Adrian Franklin (1999, 175; see also Franklin et al. 2001). In *Animals and Modern Cultures,* he argues that the social, economic, and cultural transformations that arguably constitute postmodernity "resulted in a shift from anthropocentric instrumentality to zoocentric empathy." He builds his argument in terms that hold regardless of whether you accept the premise of the "postmodern." In particular, he names three dimensions of postmodern culture that influenced this shift: ontological insecurity, risk reflexivity, and misanthropy. He then marshals evidence from several sites of human–animal interaction, including pet keeping, hunting and fishing, and the food industry, to illustrate the more compassionate attitude toward animals. I will briefly summarize the three dimensions.

If ontological *security* refers to "a sense of continuity and order in events, including those not directly within the perceptual environment of the individual" (Giddens 1991, 243), then ontological *insecurity* connotes the absence of this continuity and order. Ontological insecurity describes the sense that one's environment is unpredictable. Westerners for generations have enjoyed ontological security, but recent decades have changed that. For instance, social and economic conditions have meant growing insecurity in employment, as people can no longer count on spending years in one job or with one firm. Other examples include a lack of affordable housing and weakened or nonexistent family ties. Since the publication of Franklin's book, new examples have appeared. Corporate-accounting scandals and the attacks of September 11, 2001, have challenged the feelings of safety and security that most Americans long took for granted. In times of ontological insecurity, Franklin argues, animals offer consistency when little seems available. Evidence for this appears not only in rates of pet keeping, but also in the growing popularity of other animal-focused activities, such as birdwatching and feeding, which Franklin examines in his work.

Accompanying ontological insecurity is risk reflexivity, which refers to the realization that more than ever before, humans and non-humans share the same environmental threats. Moreover, the threats are largely due to human infiltration of every corner of what used to be wilderness. The animals within those once isolated spaces now depend on human intervention for their very survival. The variety of environmental movements and organizations to preserve specific species, such as the whale and the dolphin, reflect awareness of human responsibility. Other scholars have recognized this, as well. For example, Myers (1998, 46) writes that "we are somewhat freer today to acknowledge our biological continuity with other organisms and certainly aware it would be wise to respect our interdependence."

Related to risk reflexivity and ontological insecurity is a third, related condition of postmodernity: misanthropy. Franklin intends this to mean a general antipathy to humans as a species rather than a dislike of people or specific individuals. Misanthropy, in this sense, depicts humanity as "out of control, deranged, sick or insane" (Franklin 1999, 54) in light of the devastation wreaked on animals and the environment in the interest of progress. Misanthropy casts humanity in a distinctly negative light, in contrast to the "essential goodness, sanity, and healthiness of animals" (Franklin 1999, 55). A good example of this is the favorable, often sweet portrayal of animals that spread rapidly through films and popular literature in the post–World War II era, with Disney cartoons being the most familiar example.[3] Combined with ontological insecurity and risk reflexivity, misanthropy encourages relationships with animals as a "moral counterbalance" to the instability and immorality of human beings (Franklin 1999, 55).

To Franklin's three dimensions I would add another–specifically, the increasing awareness that animals are conscious, sentient beings. Thanks to the efforts of scientists who were willing to think "outside the box," a more enlightened perspective on animals became available. An indication came in 1973, when Konrad Lorenz (known especially for his studies of imprinting among geese), Niko Tinbergen (known as the "curious naturalist"), and Karl von Frisch (known for his work on the language of bees) received the Nobel Prize. As Bekoff (2002, 34) explains, "Many people were aghast that scientists who spent their time 'just watching animals' could out-compete biomedical researchers." It was a watershed for the field of ethology, or the study of animal behavior. A comprehensive history of this field is far beyond the scope of this book, but I will sketch out enough to make my point. Darwin (and some of his contemporaries) emphasized mental continuity across species and thus attributed cognitive states to non-human animals. His perspective is called "anecdotal cognitivism" (Jamieson and Bekoff 1993), because he made observations of particular cases and did not conduct controlled experiments. Twentieth-century efforts to make the study of animal behavior more rigorous ushered in the age of behaviorism, the various forms of which dismissed non-observable mental states in favor of rule-governed, stimulus-and-response models. Meanwhile, the notion of cognition among animals remained alive in classical ethology, represented by such figures as Lorenz, Tinbergen, and von Frisch. Then, in a 1976 book titled *The Question of Animal Awareness,* Donald Griffin rekindled interest in the idea that conscious thought was the basis for much complex behavior among animals (see also Griffin 1992).[4] Griffin founded the field of cognitive ethology, which examines the mental experiences of animals, especially as they occur in the animals' normal lives in natural settings, rather than in labs. The field shares aspects of classical ethology, comparative psychology, and the philosophy of science.[5] The perspective informs countless studies of animal cognition and emotions, including Irene Pepperberg's work with Alex the parrot (see Pepperberg 1991), Francine (Penny) Patterson's research with Koko the gorilla (see Patterson and Linden 1981), Goodall's fieldwork with chimps in Gombe, and Bekoff's work on the social lives of canids. It would also open a path for works such as Elizabeth Marshall Thomas's *Social Lives of Dogs* (2000) and *Tribe of Tiger* (1994); Jeffrey Moussaieff Masson's *Dogs*

Never Lie about Love (1997); and Masson and Susan McCarthy's bestseller, *When Elephants Weep* (1995).

Few lay people would know the term "cognitive ethology," but many are surely familiar with its claims and its orienting perspective. Just as other new ideas such as evolutionary theory percolated through culture, so has the idea that animals have mental states and emotions and that we share more than we realize. Combined with Franklin's dimensions of ontological insecurity, misanthropy, and risk reflexivity, it is easy to see how people could come to call animals companions rather than pets. Once you consider an animal as having a subjective self and the capacity to share intentions and emotional states, it becomes harder to think of him or her as existing solely for your amusement, use, and pleasure. As some people began to think in these new terms, they also needed "a word less freighted than 'owner' to describe their relationship" (Derr 1997, 324). They began to use the term "guardian" to describe the different set of responsibilities. Several cities implemented this language in their animal ordinances. The first was San Francisco, where Elliot Katz, a veterinarian and president of the organization In Defense of Animals, who spearheaded the change, called it part of a "revolution" in how people see their cats and dogs. In July 2000, Boulder, Colorado, passed an ordinance replacing all references to "owner" with the word "guardian" in city codes.

Some argue that calling an animal a companion rather than a pet does nothing as long as they remain defined as property, an issue I return to in the conclusion. For instance, although the city of Boulder has implemented the word "guardian," its website clarifies that the words "guardian" and "owner" nevertheless mean the same thing. Gary Francione, professor of law at Rutgers University School of Law, argues that language has no legal bearing because the law considers animals the property of an owner, no matter what we call our relationship with them (Francione 1995, 1996, 2000). However, others use the word even though they know that it does not legally characterize their relationship with their animals. Mark Derr, for example, writes, "Like many people I prefer to think of myself as the custodian or guardian of my dogs. I have a responsibility to provide medical care, food, shelter, and certain instruction on how to get along in the world, and legally, I own them. But morally, I do not" (Derr 1997, 324). Similarly, the holistic veterinarian and author Allen Schoen (2001, 9) explains that he uses "animal companion" and "human companion" instead of "pet" and "owner" because he does not think that "the complex relationship between two species can be rendered accurately by two such loaded words."

ANIMAL SELFHOOD AND ANIMAL CAPITAL

Schoen's words "the complex relationship" call attention to the key difference between having a companion animal and having a pet. We cannot have a terribly complex relationship with a pet. A pet exists for pleasure, and owners need only sit back and enjoy. Complexity should not enter the picture, and if it does, pet owners will usually push it out again. As one man told me in an interview:

> *I have* pets. *My dogs are pets. I don't think they have very advanced thought processes. If one of them has an expensive medical problem or gets sick, I'd just have it put to sleep and get a new one.*

Notice that a dog is a genderless object, or an "it," to him. In contrast, a complex relationship implies that we must come to know a great deal about the other being with whom we share the relationship. If we consider the relationship complex, then our interaction will require a commitment to learning how the dog or cat sees the world and functions within it. In turn, this presupposes that animals have minds and feelings that help them to know and function. This has to spark curiosity—at least, it seems to me that it must—about how animals communicate, feel, and learn, and about what they need to live healthy lives. The knowledge that comes from such curiosity produces an entirely different kind of relationship with animals.

This type of knowledge involves what I call animal capital. The term "capital" refers to resources that have currency within a given social field. People can exchange capital for advantages in particular settings. For example, to a list that includes the most familiar type of capital, which is economic wealth, there is also human capital, which consists of skills and knowledge that constitute advantages in the labor market (see Becker 1975). Pierre Bourdieu (1986) differentiated among cultural capital (including knowledge, habits, and tastes that give one access to the upper classes), symbolic capital (honor and prestige), and social capital (social networks). James Côté and Charles Levine (2002) add identity capital, which includes critical thinking abilities, a sense of purpose, and other agentic capacities. Different social fields value different types of capital. The business world, for instance, values economic and social capital, while the academic community trades primarily in cultural capital. I use the term "animal capital" to refer to resources that enable the development of meaningful, nonexploitive companionship with animals. I include knowledge about behavior, nutrition, health, history, breed characteristics, training, and the variety of things that can enrich animals' lives. I also include a rapport with animals based on an active interest in their emotions, communication, and cognition. In addition, animal capital means knowing how to find things out. This implies seeking out resources such as veterinarians, trainers, and behaviorists in case of health or behavioral problems.

The idea for the concept of animal capital came to me while I was trying to understand what made some people's relationships with animals work while others' relationships failed. It became clear that two closely related factors were involved. One was a commitment to the animal as a subjective being and an active participant in the person's own life. People whose relationships with animals failed usually stopped identifying—or perhaps never did identify—the animal as a salient part of their lives and their sense of self. A second factor common to failed relationships was the inability or failure to seek and use resources to cope with a problem, whether it was the animal's behavior or their own allergies. For example, people who surrendered cats claiming they were allergic often were not even certain that they had allergies. Granted, in some situations the only thing to do is surrender the animal. However, many people did so without first consulting a vet, trainer, behaviorist, allergist, or another source of advice. In contrast, people who worked through a difficult situation came away with a greater appreciation of the animal, expanded knowledge about the resources that are available, and a sense of satisfaction about the relationship. Animal capital not only creates more complex and therefore satisfying relationships with animals; it expands the experience of self.

Animal capital has nothing to do with knowing about animals to exploit them; it has to do with knowing about them to minimize their exploitation. It has to do with recognizing the intrinsic value of animals' lives. It makes a qualitatively different kind of relationship with animals possible, just as other forms of capital make certain human relationships possible. A person who wants to understand how dogs and cats perceive their environment, for example, will have a different kind of relationship with a dog or a cat than will the person who believes animals are not very bright. The belief that animals are simply bundles of instincts is not animal capital, although recognizing the role of instinct in animal behavior certainly is. In other words, animal capital comes through attempting to understand animals' potential, and along the way it expands our potential with animals. Perhaps the best example comes from Goodall (1999, 77–78), whose approach to the chimpanzees she studied brought her animal capital that forever changed what we know about them—and about ourselves:

> In order to collect good, scientific data, one is told, it is necessary to be coldly objective. You record accurately what you see and, above all, you do not permit yourself to have any empathy with your subjects. Fortunately, I did not know that during the early months at Gombe. A great deal of my understanding of these intelligent beings was built up just because I felt such empathy with them. Once you know why something happens, you can test your interpretation as rigorously as you like.

As she makes clear, her insight into the chimps' lived experience was far deeper because she wanted to understand them. She did not assume, as did many researchers of the time, that she had to leave her feelings behind when she made her observations, or that the chimps could have been better studied in a lab. In the context of everyday life with dogs and cats, this attitude can change the way we think about them.

Let me now anticipate a question.

ISN'T THIS ANTHROPOMORPHIZING?

It is. When I say that dogs and cats have selves and feel emotions, I am using human terms to describe animal behavior, and that is anthropomorphic. I find it so telling that, of all the different kinds of language and imagery that people have used to depict animal behavior, from that of competitive economics to machinery, the emotional and psychological descriptions arouse the most heated objections.[6] Yet, as humans, we cannot *avoid* anthropomorphizing. Kenneth Shapiro (1997, 294) points out that it is not something we do only when we try to understand animals. Rather,

> all understanding is anthropomorphic (from *anthropo,* meaning 'man,' and *morphe,* 'form' or shape') for it is partly shaped by the human investigator as subject. However, since this is a perspective or "bias" inherent in all experience, it is not an occasional attributional error to which we are particularly prone when we cross species' lines. It is a condition of science which prevents it from reaching certainty and, therefore, from supporting a positivistic philosophy.

Because all human attempts to understand and describe any phenomena occur from the human point of view, the "problem" of anthropomorphism is not unique to depictions of animals. It is, instead, a consequence of language. Bekoff (2000, 21) explains that, "in order to talk about the world of other animals we have to use whatever language we speak." Because I have only human language available to me, I have no choice but to say, "Dolly seemed sad" or "Skipper likes to lie there," to describe the behavior of my companion dogs. Anyone wishing to avoid all anthropomorphic thinking would find these descriptions unacceptable, because I have no "hard" evidence of what Dolly feels and what Skipper likes. Granted, "it is especially difficult to establish, in an empirical way, that such descriptions are justified" (Fisher 1991, 71). However, if we reject the use of anthropomorphic language altogether, what can replace it? If I restrict my depiction of Dolly to a mechanical account of the movement of her facial muscles, then I leave out my awareness of the situation in which her behavior occurred. I also leave out other awareness gained through systematic interaction with Dolly. I have seen her when she seems happy, as she does when we are getting ready to go for a walk or when she is chewing a bone, chasing ducks, running through snow, and so on. Likewise, if I restrict my description of Skipper's apparent sleeping preferences to a behaviorist explanation, then again I have left out what I know through systematic interaction. I have seen Skipper try out various places to sleep and finally take one particular spot in my study. It seems reasonable to say that he likes it there.

The term "anthropomorphism" often aims to discredit someone's claims about animals.[7] It usually implies sentimentality and inaccurate projection, which I agree should be avoided. However, the situation is not one in which "the only alternatives are an unconstrained use of anthropomorphism on one hand and the total elimination of anthropomorphism on the other" (Bekoff 2002, 49–50). A middle ground involves informed, systematic interaction with and observation of an animal. Over time, this makes possible a "critical" or "interpretive" anthropomorphism (Fisher 1991; Burghardt 1998). Critical anthropomorphism aims to do for the understanding of animal life what the *Verstehen* perspective tries to capture in human life, which is to shed light on the meaningful, subjective aspects of action. It entails grounding statements about animals "in our knowledge of species' natural history, perceptual and learning capabilities, physiology, nervous system, and previous individual history" (Burghardt 1998, 72). For example, given what I know about cats, or about a particular cat, I can make reliable statements about when a cat feels contented as opposed to fearful. Cats use explicit body language, and anyone who pays close attention over time will come to understand that dilated eyes and flattened-back ears signal fear. If I ground my statements in knowledge about normal behavior, I can safely use anthropomorphic language to label it. Indeed, I have no other choice. Although I cannot know whether the cat's experience of fear is the same as mine, the label "fear" is justified.

Shapiro (1990, 1997) proposes that we can best understand animals if we begin with the knowledge that they are embodied beings and add to that equal parts of social construction and history. For example, he begins by describing the initiation of play with and by his dog, Sabaka:

> We had furnished him with a bucket, located in our playroom, in which were kept all manner of items—discarded socks, a partial upper of an L. L. Bean shoe, a rubber watermelon, dead tennis balls, and a plastic bone. Sabaka could initiate play with one of us by pulling out, say, an old sock and approaching coyly. Alternatively, we, or one of the two neighbors' dogs who often visited, could in like manner get a sock and approach him. (Shapiro 1990, 186)

Nothing seems anthropomorphic here. Shapiro goes on to depict the play routine:

> By custom, the playing field was largely limited to a couch that he could fit under but I could not (at least not without injuring myself) and sundry other combination obstacle/hideaway/out-of-bounds pieces of furniture, all in the playroom. Occasionally, the play could break out into a hallway that made a circuit around the stairwell and then back to the playroom.

Shapiro admits that, at first glance, "this activity hardly seems worth mentioning—a simple game of keep-away and chase" (Shapiro 1990, 186). On further thought, however, everything from the "invitation ritual" to the "conditions ending it" was "all quite intricate and yet easily maintained by both players. Even more subtle were the postures, feints, and deceits, the half-executed moves during what might look, to the uninitiated, like time-outs from the game." Shapiro argues that, although he can describe the game in straightforward terms of he-went-this-way-and-I-went-that, this does not do justice to what occurred. Rather, he knows Sabaka's intentions by understanding his bodily experience, and vice versa. He invites us to think of how a tennis player plans his or her next shot by watching the opponent.[8] If you remove the element of planning the next shot and focus only on the opponent's moves, this offers a snapshot of how we can know what another's body is doing. Indeed, it is truly a snapshot, for we can retain such focus for only moments at a time. Nevertheless, those moments allow us to know what an animal is experiencing—if we reflect on what he or she has access to and considers significant. In Shapiro's example, the constraints of the space, the couch, the objects, and the other body (Shapiro's) in the game all shape Sabaka's experience. It is not a conscious shaping, in that the dog cannot know, for instance, that Shapiro cannot fit under the couch. However, being under the couch is significant for Sabaka in that it is "safe." As embodied beings, we have all had similar experiences of physical safety and enclosure; surely, this is not exclusively human.

This kind of knowledge works best when informed by the other components of the three-part method: social construction and history. In this instance, social construction refers to the various discourses surrounding Sabaka as representative of the species "dog" (and to this we can add "cat"), as an "animal" and as a "pet." The point, which I raised in Chapter 1, is that we interact with animals under the influence of various preconceptions. What, for example, are "dogs" and "cats"? Are they wild creatures, designed to live free of human intervention? Are they flesh-and-blood versions of cuddly, stuffed toys? Alternatively, are they loving, devoted companions who can reduce stress and generally add to the joy of living? All these images—and others, too—are available to us. We all make use of them in our interactions with individual animals. To begin to understand an animal's experience, we must acknowledge and evaluate those images that we use. This is also the case when we consider scientific studies of animals. These are not immune to social construction. Cognitive ethology, for example, with its emphasis on animals' consciousness and versatility (Allen and Bekoff 1997), reveals one construction of animals; the behaviorist position reveals another, in which the animal is simply a black box.

Along with social construction, our attempts to understand what is in animals' minds must also include knowledge of the history of that individual animal. Just as our interaction with other humans should take into account significant events involving that person (and perhaps the larger social group of which he or she is a member), the same holds for our interaction with animals. For example, in understanding Skipper, I have benefited from knowing that he was picked up as a stray and spent several months in at least two shelters before I adopted him. That history has helped me understand some of his behavior, especially in the early days of our life together. It also helps me to remember that he is part herding dog, which gives him particular behavioral tendencies. Similarly, in trying to understand why one of our cats runs and hides whenever strangers come to our house, it helps to know that she was born in a feral cat colony and did not have contact with humans until she was about six months old.

An important precaution is required when we draw on individual animals' histories. Shapiro puts it well:

> One implication of the recognition of the importance of an individual's history is that it undercuts an emphasis on the status of any study's findings as universal. If we take seriously the importance of individual history, we cannot claim to be analyzing the essential dogginess of the dog. (Shapiro 1990, 194)

Although I cannot generalize claims about individual dogs or cats, the same applies to claims about other people. However, there is a point at which we have enough certainty about what another is thinking or feeling to close the knowledge gap with a leap of faith. That point usually indicates something common to the behavior rather than a quirk of the explanation. For example, in the field of social psychology, as in many fields, if a number of observers offer the same interpretation of a phenomenon, then scholars conclude that something in the structure of the phenomenon produced the consistent interpretations. In other words, the consistency of the explanations must stem from the structure of that which is being observed. A recent experiment set out to test this with explanations of canine behavior. To what extent would people use anthropomorphic descriptions, and when they did so, how consistent would these be across subjects? To answer this, people watched videotapes showing interactions between dogs and their guardians and then described what was happening, focusing on the dog. They most often used psychological descriptions, and "there was remarkable agreement amongst participants in the meanings of the anthropomorphic descriptions they applied" (Morris et al. 2000, 162). In other words, something in the dogs' behavior suggested the psychological descriptions. They did not spring from ungrounded, sentimental anthropomorphism. In the study of human behavior, such findings constitute evidence of "something in the structure of human actions and posture specific to different intentions and emotions" (Morris et al. 2000, 162). I argue that this is also the case for animals.

Sentimental Anthropomorphism

In contrast to critical anthropomorphism, people often ascribe ideas and preferences to animals with little basis for doing so. For example, I have heard many people claim that when you adopt animals from a shelter, they "know" they have been rescued and feel "grateful" for the rest of their lives. For a dog or cat to understand having been rescued, he or she would need concepts for the experience of getting lost or being abandoned. These are human concepts. A dog might indeed experience fear or confusion when separated from the (human) pack, and I have seen lost dogs excited to see their guardians after having spent a day or two at The Shelter. However, I have no evidence that dogs understand concepts such as "lost," "abandoned," or "rescued." Moreover, a stray dog adopted from a shelter is no less likely to wander off from his new home than any other dog, given motivation and opportunity. So much for gratitude.

The harm in uncritical, sentimental anthropomorphism comes when it short-circuits the effort to understand animals' realities in animals' terms. For example, the claim that rescued animals are forever grateful reveals more about what the person saying this expects from a relationship with an animal than it does about animal behavior. This person wants acknowledgment for rescuing an animal, and the statement always leaves me wondering how a dog or cat would ever adequately express his or her gratitude. One possible answer for dogs is "loyalty," a common, sentimental anthropomorphic projection that has produced the notion of the "one-person dog." This is a tempting way to account for a dog's bond with a particular person, and there are times that even I want to believe it is true. Skipper will listen to me rather than most other people, and few things do my heart as much good as seeing him come running toward me at a clip when I call his name. However, having adopted him after he was a stray, I know I am not his first or his only human companion. Together, we built the relationship with time, consistency, and rewards. If something were to happen to me, Skipper could, I know, bond with another person, as much as I would like to think otherwise. Perin has explained this well:

> Dogs are not constituted to be "utterly devoted" to any one of us, in plain fact. *We* have that interest. Dogs keep on telling us that our definition of "fidelity" is not theirs when they so readily take to a new family or wander away without so much as a tear. The "one-person dog" is our myth (an observation dog haters often make). These are *human* ideas defining the dog's relationships to us. They are cultural ideas. (Perin 1981, 80–81; emphasis in the original)

The better alternative to thinking in terms of the "one-person dog" would be to understand how dogs bond with other beings. It is not sentimental anthropomorphism to say that dogs "get" hierarchies and the idea of a leader. It is not sentimental anthropomorphism to say that they learn through reinforcement, doing what they find rewarding. Granted, I must use human language to say these things, but it is not simply projection to do so.

Sentimental anthropomorphism not only does animals a disservice; it often does them harm. This is often the case with puppies and kittens, whose "cuteness" triggers a complete denial of the needs of small or young animals and the responsibility of guardianship. The full-on projection of human characteristics ignores the realities of teething, housebreaking, training, and more. One species that is a frequent recipient—"victim" may be the better word—is the rabbit. Given the little that people know about the behavior of dogs and cats, most know even less about rabbits. As a result, people often neglect them or surrender them to shelters after they reach sexual maturity and become "real" animals rather than cute playthings. More dangerous to rabbits is when people simply set them loose somewhere, mistakenly thinking that a domestic rabbit is the same as his or her wild relatives. This is another example of the damage inflicted when we fail to understand animals' lives in their terms. Anthropomorphic projection can do considerable harm when it produces a romanticized image of what animals must "really" be like, which is usually some combination of "wild" and "free." In this construal, dogs and cats are everything the modern, urban, technological world is not. At The Shelter, I often heard people manifest this by saying, "I wish I had a farm so I could just give them all a place to live." This reflects the belief that all dogs need is a place to run, and that cats should go outside. On the contrary, most dogs usually prefer to spend a great deal of time with their human pack—wherever they are—and cats live longer, healthier lives indoors. The worst consequences of the "wild" and "free" image of animals are the numbers of unwanted litters. A typical example of romantic anthropomorphism is the behavior of a man whose intact male dog was a "frequent flyer" at The Shelter, brought in repeatedly after being found stray. The man would not neuter the dog even though it would decrease his urge to wander. At the mere suggestion of it, he cringed and said, "I would *never* do that to a dog. He wouldn't know who he was."

The concept of animal capital is anthropomorphic because it portrays animals as like us in many ways. In particular, it assumes that they have minds and feelings and that they communicate in ways that we can often understand and share. It also presupposes that animals have selves. Animal capital is anthropo*morphic*, but it is not anthropo*centric*. It challenges the ideology of human dominion over animals with the possibility of cooperation. Against this, some would argue, as Tester (1992) has, that meaningful relationships with animals are impossible because animals are both subjugated to humans and unable to express their view of the relationship. However, there is growing evidence that people and animals can learn to communicate on terms not established strictly by humans. To be sure, we do not use animal capital with all animals or in all instances. Many people who recoil at the idea of using force to train a companion dog do not hesitate to eat eggs that come from hens who suffer greatly to produce them. Attitudes toward animals still vary widely and are fraught with ambivalence (see Kellert 1994; Arluke and Sanders 1996). While some species live comfortably in human households, the majority remain exploited and tortured for food, hide, entertainment, sport, and research.

In this chapter and the previous one, I have highlighted major developments in the human stance toward animals, albeit in broad sweeps. Now I will briefly summarize the discussion. Ancient human societies created and enforced a boundary separating us from animals to justify dominion over nature. The animals who helped us rather poorly equipped humans achieve dominion occupied a special position on the "borderlands" of the boundary. Throughout much of Western history, avid debate has surrounded the status of animals, the terms of which have mirrored salient moral questions of an era, including the meaning of rationality, the reach of the law, and the issue of slavery. As animals became less of an economic necessity and nature became less of a threat, the presence of dogs and cats in some human circles was justified with the term "pet." Pets were first available mainly to the elite, whose status protected them from the criticism of associating with animals. Pet keeping gradually became democratized, although evidence suggests that animals served as indicators of social class. Today, as more people come to recognize our similarities with animals, some characterize their relationships with them as companionship, although the law defines their role differently. The relationship of companion animal and guardian could not have developed at any other time in history. It hinges on the portrayal of animals as possessing cognition, emotions, and other components of selfhood. It also hinges on the accumulation and exchange of animal capital, which is knowledge, skills, access to resources, and a motivating interest in the well-being of animals. This in itself is contingent on the view that they are minded, sentient participants in interaction. As long as animals *were* capital, we could not have animal capital in the sense that I use it here. Although there were surely individual instances of closeness between people and animals, until recently it would have posed too great a threat to the human–animal boundary to consider dogs and cats companions rather than pets. Just as the pet became possible only when conditions protected some people from the stigma of associating with animals, the companion animal became possible only when we began to recognize our continuity with other species and our obligations to them.

4

Looking at Animals/ Glimpses of Selves

Perhaps we are looking for self in all the wrong places or faces.

—Marc Bekoff (2002, 98)

One of the most contested implications of our continuity with other species is the possibility of animal selfhood. I argue that animals help to shape our identities because they bring selves of their own to the interaction. As most people who live with animals know, animal selfhood becomes present in long-term human–animal relationships. However, it is apparent even in brief and periodic interactions, such as those I observed among people visiting homeless animals at The Shelter. These clients constitute a category distinct from those who intend to adopt, and I examine the latter group's interaction with animals in the following chapter. Regardless of intention to adopt, clients who come to The Shelter may walk freely through the adoption areas, see the cats and dogs, and read whatever information is available about them. I became curious about why scores of people would want to look at homeless animals, given that they had no plans to give any of them a home. The answers open the door to the discussion of selfhood—both human and non-human—that continues throughout this book.

Before I get to the heart of things, let me first set the scene by describing what people encounter when they visit The Shelter. As I just mentioned, the adoption areas are open to the public. On an average day, about thirty dogs and sixty cats await adoption, but the number of cats can grow much higher during "kitten season," roughly May through October. The dog-adoption area contains kennels; glassed-in "apartments," which can house two dogs or a litter of puppies; and a "playroom," which houses several dogs at once. In the cat area, individual kennels accommodate individual cats, and several "apartments" hold between three and five sociable cats. The cat "apartments" have shelves for sleeping and observing, as well as toys and climbing equipment. When the number of cats exceeds the available spaces, the dogs' "playroom" is used for cats instead.

A card on each kennel tells the animal's sex (all animals are sterilized); age (or an estimate); breed (or an educated guess at the mix); means through which the animal arrived at The Shelter (as a stray, transferred from another shelter, or surrendered by a guardian); and other pertinent aspects of his or her history. The kennel cards also list the animal's name—or, in the case of strays, the *current* name. The amount of information available about an animal varies widely. When guardians surrender animals, Shelter staff asks them to provide basic information along with a reason for surrender (for example, moving, allergies, etc.). In the best cases, guardians provide other details, such as the animals' favorite food, toys, and activities so that a potential adopter can learn about habits or preferences the animal has. However, many guardians provide little of this, and animals coming in as strays bring none with them. The kennel cards also have space for additional, handwritten comments, often added by volunteers such as me who become familiar with the dog or cat. This section can include jottings such as, "Rides well in the car," "Great Frisbee dog," or "Loves to be rubbed behind her ears."

When a potential adopter wants to meet an animal, he or she brings the kennel card to the front desk. An adoption counselor uses its information during the introduction and, if no adoption occurs, puts the kennel card in a bin for replacement on the kennel as soon as possible. On busy days, however, "as soon as possible" is not always right away, and consequently some animals' kennels are temporarily without information.

"JUST VISITING"

The majority of Shelter clients come solely to *see* the animals, *not* to adopt them. This is not to say that none of these clients ever adopts. Rather, it means that most people strolling through The Shelter at any given time do not intend to adopt animals at that time. When I approached clients who were strolling through the adoption areas and asked whether I could help them, they answered, "No, we're just visiting," much in the same way I remember my mother saying, "No, thank you; just browsing" as we wandered through stores in which we had no intention to buy. Consequently, I call this category of interaction "just visiting" and the clients "visitors."

Many visitors bring their young children to see the animals, and their trip to The Shelter is a planned outing, much as a trip to the zoo would be. For many children, The Shelter offers a first encounter with the idea of homeless animals. In a typical interaction, a child, on seeing the kenneled dogs and cats and seeing me there, too, would think that the animals were mine. "Where did you get all these animals?" they would ask. "They came here because they had nowhere to live," I would answer, "and we'll take care of them until they find new families." That would begin a series of "why" questions, the course of which will be familiar to all parents: Why don't they have homes? Why can't they stay here? Why can't they live with us? Why do they have to be in cages? Some visitors, with or without children, asked to touch or hold certain animals and spent time with them outside their kennels. Many already had a houseful of animals. Nevertheless, they came to look. A number were "regulars," stopping in so often that I came to know their names. Some visitors came to check on animals they particularly liked, especially those who seemed to be having a hard time finding a home.

The numbers of visitors compared with the number of people who actually adopted animals came as a surprise to me when I first began volunteering at The Shelter. How to turn "visitors" into adopters is a vital concern among animal shelters. At almost any given moment, a dozen people are wandering through the dog- or cat-adoption areas "just visiting." Similarly, only about 2 percent of the clients who board the Mobile Adoption Unit adopt animals. Granted, people *do* adopt, and I analyze that decision in the next chapter. The point I want to emphasize here is that most people want only to "visit" the animals, and the appeal and meaning of this as an activity deserves investigation.

POSSIBLE SELVES

Because I have already compared looking at animals to window-shopping, I will follow the implications of that comparison. As John Gagnon (1992) argues, window-shopping presents opportunities to fantasize.[1] Shop windows, like advertisements, exist to produce desire. By looking at objects on display, we can fantasize about what it would be like to have them. *What might happen to me if I wore that red sweater? That hat? If I had that car? Wouldn't it be great if I could wear those shoes to the party?* The merchandise evokes different possible selves. We have a similar experience when we walk by a restaurant and see the patrons virtually "on display" eating and drinking. *What would it be like to sit there and eat that? Look at them, enjoying their margaritas. I want to do that, too.* Like window-shopping, looking at animals offers new opportunities and new possibilities for selfhood. *Remember that cocker spaniel I used to have? What would it be like to have a dog again? I wonder what would it be like to have* that *dog?* Those *kittens? What would I be like if I had a black cat?*

Visitors often came in pairs or groups, and those who did so spent more time talking to one another than looking at the animals, something noted in other studies of shelter clients' behavior (see Wells and Hepper 2001). This suggests another way in which animals serve as "social facilitators," as I discussed in Chapter 1. The conversations that took place in the adoption area show that a considerable amount of "trying on" of possible selves takes place, as these snippets reveal:

> *If I had a dog like that, I'd have to start running again.*
>
> *I could never have a dog with long hair. I can't imagine having to vacuum up all that fur.*
>
> *We used to have a cat just like this when I was a kid. When I have my own place, I'll get another one.*
>
> *Remember when we had Toby? Here's a dog just like him. I wonder if he likes to swim, too.*

At least part of the appeal of strolling through The Shelter with no intention to adopt is that it offers a temporary escape, much like shopping or going to the movies, detaching the person from who he or she currently is.

That visitors intended only to "try on" potential animal-related selves became increasingly evident in light of the disclaimers they offered for just visiting: "I can't adopt any, but I just had to visit"; "My husband will kill me if I bring any more animals home." In a social-psychological sense, the use of disclaimers shows that the person using them understands in advance that his or her actions could be discrediting (see Hewitt and Stokes 1975). Disclaimers ward off the potentially negative implications of what someone is about to do or say (see Hewitt 2000). We make use of them when we say such things as, "This might be a stupid question, but . . ." or "Don't get me wrong, but. . . ." Inserting the disclaimer acknowledges the risk that we will be misunderstood and attempts to set things right. Thus, when people looking at homeless animals say, "I already have three cats. I can't afford to feed any more," they are acknowledging that their actions occur in a context in which others may legitimately hold some hope or expectation that they will adopt. Visitors "disclaim" their actions to avoid being typified as potential adopters. Their use of disclaimers implies that they do not mean to mislead anyone. Voicing the disclaimer in advance establishes their definition of the situation as the "real" one. They assert that they intend only to look at the animals, not to go home with one.

"JUST VISITING" AND EMOTIONS

There are important differences between imagining how a new pair of shoes could change one's perspective on life and fantasizing about living with a cat or dog. The animals at The Shelter, unlike the shoes, the sweater, or the hat, are living beings who desperately need new homes. Indeed, some people avoid shelters because they find seeing all the abandoned and neglected animals too difficult, even in an environment such as The Shelter, where the animals all eventually will find new human families. However, because so many people do enjoy looking at the animals despite the emotional wear-and-tear of doing so, the emotional experience may be a relevant part of the meaning of the activity. Indeed, "just looking" may be enjoyable *because* of the wear and tear. Two distinct emotional experiences are available to visitors.

The first has to do with the sense of mystery and yet familiarity that animals evoke. John Berger (1980) examines this in the context of looking at captive wild animals in zoos, but his observations pertain to other contexts, as well. Berger argues that, through looking at animals, people encounter the sense of simultaneous similarity and difference that has always existed between "us" and "them" and that remains a constituent part of our existence. Humans were once in frequent, close contact with animals; they were the first symbols, providing metaphorical answers to our questions about how the world began. As our own species gradually came to dominate the world, we marginalized animals. Now most people know animals mainly as pets, cartoon characters, stuffed toys, or dinner. Yet, according to Berger, something of the original, mysterious connection remains. We can keep animals in captivity and still not understand them. When we look at them, we experience the ancient sense of "like us" and yet "not like us," and we sense the power we have over these other beings. Zoos remain popular tourist attractions, Berger argues, drawing greater numbers than most professional sporting events because they are sentimental monuments to the superiority of humans.[2]

Although the element of "the animal as spectacle" exists more clearly in the case of the captive wild animals, the same phenomenon nevertheless occurs in shelters. To the extent that the animals are "on display" in cages, shelters are like zoos—with one important difference. In zoos, the animals are wild, coming from faraway, "exotic" locales, whereas in shelters, the animals are familiar. They are the dogs and cats who normally live with humans and yet are, in the context of the shelter, also "displayed" in cages. Although they are not the species we find in zoos, they nevertheless "stand in for all members of the animal kingdom, permitting a distant, ritualized contact with other species" (Melson 2001, 29). Looking at animals, as Berger argues, has always brought a mixture of emotions, including wonder, superiority, and envy. Just visiting homeless animals offers people a version of this age-old experience.

A second set of emotions comes from awareness of the animals' circumstances. For some of the visitors I spoke with, the emotional to-and-fro seemed part of the purpose of the visit. For instance, of the responses I recorded while on the adoption unit, nearly one-third of the visitors said things like the following:

> *It tears me up to see them, but I just have to see who you have with you today.*
>
> *It breaks my heart. I wish I could bring them all home.*
>
> *I can't stand to see them in cages, but I have to say hello anyway.*
>
> *I don't know why I do this. It's so depressing. I couldn't stand to do what you do.*
>
> *I could never do what you do. I'd want to bring them all home with me.*

Clearly, visiting homeless animals produces a complex mixture of feelings, but in light of these comments, that experience seems part of the reason for the visit. For adults, elements of the pleasure of looking at animals surely endure from childhood, but the pleasure mixes with harsher emotions brought by knowledge of the reality of the environment. The visitors' comments indicate that seeing the animals reminds them of the irresponsible ways we treat our closest companions. At the same time, as the last comment reveals, some of the guilt and heartsickness over our collective neglect is tempered by the knowledge that—in a shelter, at least—someone is caring for these animals, albeit temporarily.

The mixed emotions give rise to an emotional-distancing device that I call the "I could never" strategy. I think of this as the personal version of the phrase, "Not in my backyard," which people use to evade responsibility for environmental hazards. With "I could never," people evade responsibility for hazards of an emotional sort. First, the response allows people to take the moral high ground over others. "How could anyone give this animal up?" I have heard countless people say. "I could never do that. What kind of person could do such a thing?" A second "I could never" usually follows the first, this one directed at staff and volunteers. "I could never do what you do. It's just too depressing." With this second "I could never," they imply that, because they are so sensitive to the plight of homeless animals, they would find working in a shelter too hard (see also Arluke and Sanders 1996, esp. chap. 4; Rollin and Rollin 2001). If a *sensitive* person could *never* do the job, then shelter workers must somehow be immune to the emotional hardships. In reality, this is clearly not the case. Workers must balance caring and compassion with anger and grief (see Brestrup 1997; Irvine 2002), and the job comes with a high rate of burn out (see Rollin and Rollin 2001). The "I could never" emotional-distancing device allows for the pleasure of looking at animals while evading responsibility for their abandonment and subsequent care. In many ways, this device also allows for the trying on of "possible selves," because it poses certain actions as out of the question, given the presumed self of the speaker.

This kind of experience is not available from window-shopping. Shoes and sweaters do not generally evoke emotional-distancing devices. Although I might have to manage emotions in the context of wanting some object or merchandise, the act of convincing myself that I do not really need new shoes is different from that of convincing myself that

I am not at least partly responsible for the overpopulation and neglect of animals. The complex emotions of wonder, guilt, superiority, and awe come because the animals are living beings with their own histories, preferences, and needs. In other words, just visiting brings a distinct emotional experience because the animals have selves, and because interaction with them evokes and confirms our selves.

GLIMPSES OF ANIMAL SELVES

The typical visitor will walk past most of the kennels and stop to look at only a few animals. Of approximately twenty dogs, for example, he or she might interact with three. Among cats, the rate is even lower. The Shelter usually houses thirty to fifty cats, and visitors might spend time with only one or two. In addition, visitors who came with others or in groups of three or more spent less time interacting with the animals than did those who came alone (see also Wells and Hepper 2001). This reveals how little time most people spend actually looking at the animals they ostensibly came to see. More important, it reveals that people really came to see *certain* animals, although what constitutes that category of animal would become clear to them only in the act of seeing the entire selection. This shows that surprisingly few animals are actually "contenders" for adoption at any given time.

Other studies corroborate this. Videotapes of clients' behavior at one dog shelter reveal that they express interest in only a small proportion—fewer than 30 percent—of the available dogs (Wells and Hepper 2001). "Something" about particular animals attracts particular people. It may be a certain breed or type, but it is not always that specific. However, as I will show in the next chapter, people are often attracted to animals who are exceptions to what they would list as "ideal" for them, physically or behaviorally. There is a "right" animal for each person, and vice versa (see also Alger and Alger 2003). This rightness becomes apparent (or not) as elements of the animal's core self become present to the visitor, and this foreshadows the analysis that will run through the following chapters.

I saw other evidence that people had a tacit preference for particular animals. Visitors who wanted to touch and hold certain animals were very clear about wanting to do so only with that animal. In other words, they wanted to hold a particular cat, not just any cat. For example, a visitor looking at a kennel housing six kittens, all litter mates and three of whom looked very much alike, wanted to hold a specific one. When I picked up the kitten to whom I thought she had pointed, she called out, "No; not that one. The *other* one. The one with the red collar." Because she did not intend to adopt—she just wanted to hold a kitten—it would seem to make little difference. However, that she saw the one kitten as different from the others suggests that there was something inimitable about the chosen kitten. The woman could not simply project the features of the chosen kitten onto another. Again, I argue that elements of the kitten's core self became available to the woman—available and appealing, for reasons we shall see.

These instances call attention to how animals exercise and challenge our emotions and our interactional skills in ways that are both like and unlike our interaction with other people. I have said that looking at animals allows for the "trying on" of possible selves. The presence of an animal in our lives—or even the imagined presence—evokes other potential ways of being. Because animals have unique histories and needs, even imagining them in our lives changes us. Their presence would produce a new structure for our days, as we imagine the necessary feeding and walking or running, the naps, the affection, and everything else that a new companion might bring. In addition, because their histories and needs are so evident in a shelter environment, looking at them offers an emotional experience that is not available in most other contexts. I can think of no other activity that arouses such a complex mixture of feelings as visiting homeless animals. Without uttering a word, animals can make people feel guilty. Moreover, without uttering a word, they can make us feel delight, wonder, and a host of other feelings. In addition, animals, like the people in our lives, are not interchangeable. One kitten is not the same as another kitten, because we act and feel differently about that one kitten. One kitten will engage us, surprise us, and generally interact with us in ways that no other kitten will.

Our continuity with and obligations to other animals involves recognizing that animals are self-aware, perhaps not in the same ways that humans are, but in all the ways that they need to be as animals (see also Bekoff 2002, esp. chap. 4). The interaction in the adoption area among those who are "just visiting" offers glimpses of animal selfhood. It becomes apparent even through kennel doors and with limited contact. If animals were merely the byproducts of anthropomorphic projection, then any cat or dog would provide what visitors seek. That is not the case. If visiting the animals were just like window-shopping, then doing so would not arouse such intense emotions. Because animals are not cars or clothing, we experience them as selves in relation to our own, much like the experience of interacting with other people. Myers (1998) refers to this as "animate relating." Elements of animals' selves become present to us through interaction, even in the relatively limited context of a visit to The Shelter. This becomes more evident in the experience of people who do intend to adopt, which is the topic of the next chapter.

5

The Adopters: Making a Match

Only Connect.

—E. M. Forster, *Howard's End* (1921, 214)

Here is a description, drawn from my field notes, of one "type" of potential adopter:

I have in my hand a list of dogs to photograph for the website when The Shelter opens for the day and the first clients arrive. One of them is a woman who has been looking for just the right dog. She adopted a cocker spaniel mix about a year ago. At the time, she had an elderly beagle, but the dog since died and she now wants a second dog as company for the spaniel. She wants another beagle, between two and four years old. From our brief interactions, I know that she looks at The Shelter's website regularly and then comes in when she sees a dog who interests her. She has consequently met several beagles and beagle mixes, but not the right one. Today she has stopped in because she saw that we have another beagle. The dog is out on a walk, I tell her, and she wants to wait to meet him even though, at eight years of age, he is much older than she really wants. Coincidentally, the next dog on my list is also a beagle, but a female, about three. Because the dog does not yet have her picture on the website, the woman has no idea that she is available. As I lead the new dog out of her kennel for her photo session, I ask the woman if she would like to meet her. "Sure," she says, and smiles broadly. I must admit that, although I prefer larger dogs, this little one is quite a character. As the woman bends over to pet her, the dog wiggles her entire back end with pure delight, tap-dancing with her front paws. She has bright brown eyes, and the woman remarks on how much the dog resembles her dog who recently died. I take the woman and the dog to a "meeting room" to be acquainted. We crouch together on the floor, and the dog begins to put on her charm. Once again, the woman says how much the dog looks like the one who died. I look at the kennel card and see that this dog was a stray. I tell the woman where the animal-control officers found the dog. We both shake our heads and discuss for a few minutes how this little dog survived in the busy area. I tell her that we really do not know much about the dog since she was a stray but remind her that staff members have done a temperament evaluation and a health check. As I look over the paperwork, I relay the findings to the woman. The dog has no obvious health problems but seems a bit nervous, so she would be best in a home without very small children who might over-excite her, which could lead to nipping. This suits the woman because her grandchildren are older. The Shelter's vets had spayed the dog three days before, and I explain that she needs to stay dry and reasonably calm for about another week. I explain that the dog has absorbable sutures. Then, as if she heard me, the dog rolls over on her back to display them—and to get a belly rub, which further reminds the woman of her deceased dog. The woman asks the dog in a soft singsong, "What were you doing out wandering around? Do you want to come home with me?" The dog then turns over, stands on her hind legs, and places her front paws on the woman's thigh. The woman brings her face close so that the dog can lick her. The woman laughs and talks softly to the dog for a minute, saying, "Would you like to come home with me? Do you think we can be friends?" All this time, the

> *woman is petting the dog and caressing her long ears. She finally stands up, looks at me, and says, "You don't have to take her picture. I think I found my dog." "Great!" I say and put the beagle back in her kennel to wait while the woman completes the adoption paperwork.*

For comparison, a man describes a different experience:

> *I had just moved into this place where I could have a dog, and it's been a few years since I've had one. I've had dogs all my life until just a few years ago, and so I went to The Shelter to get one. I had nothing particular in mind. I just wanted a good dog. Breed didn't matter. Age—not old but not a puppy, either. I just wanted a nice dog. I looked at a couple, and this one, we just connected.*

Finally, here is how an interviewee described meeting the cat she adopted:

> *I was with my friend, and we were going shopping. We saw the Mobile Adoption Unit in the parking lot, and I was just going to stop in to give a donation, you know, and I had no plans to adopt. I was facing away from the kennels, and all of a sudden I just had to look over at the cats. I mean, I felt this pull that I couldn't have resisted. It was almost as though he was calling me, but he was curled up asleep. The moment I saw him, he opened his eyes, and there was just something there between us. I felt like I knew him, and he knew me. Here's this not very attractive, very old cat, and I knew—I absolutely knew, beyond a shadow of a doubt—that we were supposed to be together.*

The first example illustrates interaction typical among the clients I call "planners." They come to The Shelter planning to adopt and knowing what kind of animal they want. They often have it pinned down to a specific breed, but if not, they usually have a particular size or type in mind. Some planners have all the characteristics mapped out precisely: They know the sex, size, age, and temperament of the dog or cat they will take home. Often, planners are trying to "replace" an animal who has died, and thus seek a new animal of the same type. Perhaps they just like shepherd mixes, or they have always had gray, female cats in the house. Some planners have been waiting until they have a specific type of living situation, such as when they own a house rather than rent an apartment. Even before they unpack the boxes in the new place, they head to The Shelter for the dog or cat they have waited for.

The second example involves people I call the "impartial." Like planners, they have had animals before. However, unlike planners, they have few or no explicit physical preferences sketched out beforehand. They simply hoped to find a good match and are usually open-minded about how it is packaged. The third example is of a subset of impartial adopters whom I call the "smitten." The basis for their attraction seems almost extrasensory. They describe feeling mysteriously pulled to particular animals. Sometimes it is the animal's appearance that draws them, but the pull is usually much more elusive, as in "there was just something" about a particular dog or cat. Whereas planners came armed with ideas about the animals they wanted to adopt and the impartial wanted to find a "good match," the smitten described feeling an irresistible pull from animals, even in the absence of (or in contrast to) ideas about the kinds of animals they liked.

DIMENSIONS OF ATTRACTION

The sources of information that adopters use to determine whether an animal is the "right" one are physical appearance, behavior, and a sense of "connection." Separating these dimensions to organize the discussion is an oversimplification. In reality, no adoption relies on only one of these. However, for the sake of clarity, I will first reduce the experience of meeting an animal to three dimensions.[1] Then, by the end of the chapter, I will show that the situation is not nearly so simple.

Appearance

As people walk from kennel to kennel at The Shelter, the first piece of information available about an animal is appearance. Like people, but to a greater degree, dogs and cats are embodied beings. Their physical bodies obviously reveal size, shape, and coloring immediately, but bodies reveal other things, too, such as age, demeanor, and a general sense of health—or lack of it. I found that many of the same processes that occur during attraction to another person hold for our impressions of animals. Although everyone knows that beauty is only skin deep, studies consistently show that physical attractiveness heavily influences our impressions of other people (see Dion et al. 1972; Feingold 1990). Indeed, good looks rate highest on the list of characteristics that make others attractive. People respond more positively to attractive people, whether in dating or in business situations. Moreover, people within a given culture agree strongly on what constitutes physical attractiveness, and we acquire these standards early in childhood. Granted, attraction involves more than physical appearance, but appearance is often what we judge first.

Just as there are cultural standards for attractiveness in humans—symmetrical features; large eyes; smooth, youthful skin; shiny hair—there are also standards for animals. One factor that makes animals appealing is *neoteny*, or the retention of juvenile physical and behavioral characteristics into adulthood.[2] People value youthful appearance in animals as much as they do in other humans. As a result, puppies and kittens find homes very quickly, as do adults who retain something from puppy- or kittenhood. Neoteny elicits what James Serpell (1986, 82) refers to as the "cute response." We want to hold, cuddle, and care for animals who appear young and vulnerable (see also Midgley 1983; Lawrence 1986; Melson 2001). In an evolutionary sense, the youthful appearance is a mechanism that solicits care from others, which improves an individual's chances of survival. Over time, dogs and cats who looked and behaved as juveniles received better treatment than did their wilder-looking kin, resulting in selection in favor of neoteny. At The Shelter, animals who look younger than others will usually find homes more quickly. Not surprisingly, I found that children were most often attracted to puppies and kittens and to adult dogs or cats with long hair or fluffy coats. To be sure, there are people who prefer adult and even geriatric animals; I am one of these. However, for the most part, youth—or, at least, the appearance of it—wins out. Additional characteristics that seem to make animals, especially dogs, appealing, are certain coat colors, such as gold, white, and gray. By contrast, black dogs and cats often have difficulty finding new homes.[3] In addition, and consistent with previous studies, I found that animals with a history of living indoors (indicating thorough socialization) were adopted more quickly (see Posage et al. 1998).

Planners and Appearance

Planners had a thoroughly articulated notion of the appearance they wanted in a dog or cat. Often, this had developed over years of living with a particular breed or type of animal. Consequently, for planners, appearance was often an indicator of an animal's personality: They expected animals who looked a certain way to behave as they had known another animal to behave. For example, when one planner couple wanted to adopt a cat, they knew they had to find one who would get along with their dog. They had once had a beloved brown tabby male who was what I call a "recovering tomcat," meaning that he had not been neutered until adulthood. One of the physical effects of testosterone on male cats is that it enlarges the jowls. This gives tomcats large, powerful-looking faces. Once neutered, they retain this distinctive look. Tomcats are often solid muscle, too. The couple's former cat had been able to hold his own with a friend's large, energetic dog. When the couple decided to get a cat, they chose a lush-coated, male, brown tabby with big green eyes. He, too, was a "recovering tomcat," caught in a humane trap and brought to The Shelter. The veterinarians neutered him soon after his arrival, but he had already developed characteristically male jowls and a muscular build. His appearance made the couple feel that they knew him by his resemblance to their former cat. "We liked his eyes and his face immediately," the woman told me, "especially his big cheeks. His face just said,

'I'm a tough guy–sweetie pie type,' like our old cat. We knew he could get along with our dog." His appearance even overruled negative aspects of his behavior, which was standoffish at first. As the man explained, "He was a bit reserved when we went into the room to meet him, but we liked the way he looked, so we adopted him."

Many planners seemed unaware of or embarrassed by their preferences for animals of a particular color or size. Some even asserted that appearance was decidedly not important to them. For example, they gave a rather detailed description of the traits they wanted in a dog or cat. Then, when shown an animal known to have those traits, they rejected him or her based on appearance. The most striking example of this is the prejudice toward black cats, who have a far more difficult time finding homes than cats of other colors. In a typical example, a woman who wanted to adopt a cat insisted that the most important thing was the cat's personality. Sex and age did not matter, she said, but the cat had to like other animals and could not be skittish around children. I directed her to a particular cat known to be sociable, having lived with children, a dog, and another cat in his previous home. "Oh, no!" she cried. "I couldn't have a black cat. I just couldn't have one in the house, what with it crossing my path all the time" (see also Karsh and Turner 1988). The color prejudice also holds for black dogs, although not as strongly. During my work in the adoption area, I often heard people say that black dogs looked mean. The exception to this, of course, is puppies, whose "teddy bear" appearance makes it hard for most people to imagine even the possibility of a mean temperament.

The Impartial and Appearance

In a typical example, an impartial couple adopted a dog who was a long-term resident at The Shelter, often overlooked because he was older, a bit overweight, and somewhat generic-looking. As the product of very random breeding, he looked like no breed in particular. In addition, his muzzle had some gray in it. The couple had come looking for "a medium-size, adult dog who liked people," in their own words. Having no established preferences, they found a good match that day.

The impartial often find an animal whom they like attractive, in contrast to planners, who like an animal because he or she is attractive. Moreover, the impartial often come to see an animal as attractive once they establish a relationship with him or her. As one impartial man explained, "I think any dog that I would like I would eventually come to find good-looking." The impartial are more likely to adopt animals with unique aspects to their appearances, such as one spot of black on an otherwise pure-white coat, or one blue eye. They are also more likely to consider animals with "defects," such as a missing eye or leg. Sometimes the defects and the unique appearance come in one package. For instance, before we were married, my husband, Marc, fell in love with and adopted his first cat while she was recovering from having been hit by a car. She had a head injury; she walked in circles, and it was not clear how well she could see out of one eye. I was fostering her during her rehabilitation, after her stay at the veterinary clinic but before she was well enough to go into the adoption area. Fortunately, she never had to do so. Marc met her, and her injuries made no difference. On the contrary, he looked forward to helping her with rehabilitation by stimulating her with play. The feline survivor who stole his heart is black and white, with the standard black "bandit" mask and a white face, chest, belly, and socks. But her nose is also black, in a perfect little diamond shape. That not only won Marc over; it also won the cat her distinctive name: Punim, which is Yiddish for "face." Marc never imagined having a cat, so it is little wonder that he gave no thought to how his cat would be packaged.

Appearance Matters

The adopters manifest two different orientations toward an animal's appearance. The impartial come to appreciate qualities that are not always considered traditionally attractive. It is not that appearance is entirely unimportant to them, but they have no "ideal" animal in mind. For planners, physical appearance in the conventional sense matters from the start. For both types, however, appearance matters to some degree, and I began to wonder just how it does.

One possibility is that we appreciate animals aesthetically, for lack of a better word, much as we appreciate music, artwork, or nature. I say "for lack of a better word," because strictly speaking, the term "aesthetic experience" can apply only to works of art (which is considered the modern meaning of the term) or experiences of "the beautiful" (in the traditional sense of the term, originating in the Greek work for "sensation"). Moreover, in the conventional usage, only objects to which we have no personal interest or attachment can evoke aesthetic experience. Thus, a painting of a dog or cat could produce an aesthetic experience, but our own dog or cat could not. Granted, there are differences between looking at animals and looking at art. The animal is a living, dependent creature. If a person somehow acquires the "wrong" work of art, the painting or sculpture will not suffer from neglect, whereas the animal will. For this, and perhaps other reasons, my use of "aesthetic experience" to refer to the physical appearance of animals may be a stretch, but I am not alone in extending the term to territory beyond the artistic and the beautiful. Some philosophers have suggested "widen[ing] our sensitivity to the possibilities within the notion of the aesthetic" (Diffey 1986, 10). Similarly, psychologists have argued that "the aesthetic experience is part of a larger family" of focused, rewarding, transformative experiences (Csikszentmihalyi and Robinson 1990, 9). I join them in using the term to refer to the pleasure found in animals' physical appearance, experienced as joy, delight, or wonder. The question remains: How does appearance matter?

To continue using the vocabulary of the aesthetic experience, some of the theories that apply in the context of visual art may help. These fall loosely into three categories.[4] The first emphasizes the "cognitive" dimension of the aesthetic experience. Building on idealist philosophy, the cognitive viewpoint assumes that art represents ideal forms of the things that appear to us everyday in the world. Aesthetic pleasure comes from the correspondence between a "mental model of perfection and an actual specimen" (Csikszentmihalyi and Robinson 1990, 12). Another aspect of cognitive theory emphasizes developmental processes (see Parsons 1987). For example, young children typically like realistic representations, and they equate what is good with what they like. It is usually only in adulthood that we gain the capacity to appreciate abstract art or find pleasure in a type of art we had never before considered worthwhile. A second set of theories emphasizes "sensory" aspects, suggesting that we are hardwired for aesthetic pleasure for various reasons. In an evolutionary take on things, some versions of sensory theory maintain that people who have a preference for order are better adapted to the environment and thus have a better chance of survival (see Gombrich 1960, 1979; Arnheim 1971, 1982). In another variation on the sensory theme, John Dewey (1934) maintained that the recognition of order and wholeness in art would form a model for order and wholeness in the individual and society. A third category of theories, represented by Aristotle and Freud, can be considered "cathartic." This viewpoint holds that artwork arouses strong emotions that are ordinarily stifled or denied. In arousing and subsequently purging these feelings, the aesthetic experience brings consciousness into harmony. Although catharsis is psychoanalytically feasible, there is scant empirical evidence to support it (see Csikszentmihalyi and Robinson 1990, 15).

The responses of people looking at dogs and cats split evenly between cognitive and sensory approaches. In the cognitive perspective, the pleasure in looking at an animal might come from finding one that meets a standard of what an "ideal" dog or cat should look like. Consider the exchange I had with a woman as she considered adopting a Great Dane:

> *He's so beautiful. I can't get over it. He* must *be a purebred, don't you think?*
>
> LI: *We don't know. We don't know anything about him. He was abandoned.*
>
> *I can't believe it. I've never seen a purebred except in pictures, but he* must *be a purebred. He's such a handsome dog. I've always wanted one.*
>
> LI: *He* is *handsome.*
>
> *This must be what they're* supposed *to look like [referring to the dog's uncropped ears and undocked tail]. He's* gorgeous. *I love him!*

Notice that she had never before seen a purebred Great Dane or one in the "natural" state (before the cosmetic mutilations required by the American Kennel Club). She had an "ideal," gathered from pictures in breed books, but her first encounter with the "real" version exceeded her expectations. Her genuine pleasure should be clear from her repeated exclamations ("can't believe it," "can't get over it") and her praise ("beautiful," "handsome," "gorgeous"). The text fails to convey her tone of voice, her facial expressions, and her complete absorption in looking at the dog. She was delight personified, which may have come from a cognitive correspondence between an ideal and reality.

Adopters with children gave me an opportunity to see developmental aspects of the cognitive approach. Children especially like dogs and cats who are the living representations of their stuffed animals, the characters in picture books, or movie and television animals. Whenever children were in the adoption area, depending on the selection of animals, I heard versions of the following:

> *Here's a dog who looks like Wishbone [a dog on a popular Public Television show].*
>
> *He looks like Garfield.*
>
> *This one's like the cat in that commercial.*

Adults, by contrast, can (fortunately) appreciate the three-legged dog or the one-eyed cat and can fall in love with an animal who is not "their type."

I saw evidence to support a version of the sensory perspective, too. I have already mentioned that people find symmetry appealing in other people and in animals. Indeed, men and women considered handsome or beautiful are often those with symmetrical features. The typical explanation is that we find symmetry appealing because it suggests greater reproductive fitness. This is so regardless of whether we intend to reproduce. Likewise, the baby-faced cuteness of neoteny is appealing because it arouses caregiving impulses. However, one need not accept a genetic explanation to agree with the sensory perspective. One possibility is that the impulse originated in genetics but then became part of culture. This could explain why youth and symmetry appeal widely, even among those who are not seeking "good breeding stock."

Finally, social-psychological studies reveal that, among humans, there are complex links between appearance and behavior, and between appearance and expectations of behavior. We tend to think attractive people are more competent and friendlier, and we tend to give them the benefit of the doubt more often; the benefits of physical beauty seem endless.[5] We even think more highly of people who are in the company of attractive people. This seems to work with animals, too. For example, a study of dog guardians found that women who had dogs perceived themselves as more attractive than women who did not have dogs (Serpell 1981). The women in the study may have been on the receiving end of a self-fulfilling prophecy, because how others treat us affects how we think about ourselves. If a dog or a cat thinks we are wonderful, perhaps we will think of ourselves in a better, more attractive light: how much more attractive we must be if an attractive animal thinks we are wonderful.[6]

To recap the discussion thus far, appearance matters for all adopters, regardless of whether they look for attractive animals from the start, as planners do, or for the "right" animal and then come to find that animal attractive, as the impartial do. There are two possible explanations of how appearance matters: one aesthetic, and one social psychological. The first holds that the pleasure is rewarding in and of itself, for reasons that originate in cognition and the senses. If we stretch the definition of aesthetic experience beyond the conventional use, then looking at an attractive animal may be transformative and rewarding in the same way that looking at works of art and listening to music can be. In addition, the locus of the aesthetic experience is the work of art or the piece of music itself. The art or music does not refer to anything outside or apart from it. In addition, the animal is the locus of the pleasure we find in his or her physical appearance because we experience animals' bodies immediately and directly. Humans can use clothing, jewelry, makeup, hairstyles, and other resources to accentuate, conceal, and otherwise manipulate the "raw material" of the body. Animals cannot. Moreover, humans can use spoken language to supplement or contradict what the body communicates, but animals cannot. This immediacy of the animal body leads us to locate the aesthetic pleasure that they bring solely in them, rather than in clothing and other accoutrements.

The second answer suggests that an animal's attractiveness correlates positively with desirable behavior, or the expectations of such. This, in turn, reflects well on the adopter. The animal's appearance thus brings a social-psychological payoff. Moreover, if we think an animal is intelligent and competent, we infer a sense of coherence, in that the intelligence and competence originate in the animal as a physical entity. Therefore, both answers to how appearance matters arrive at the same conclusion. Appearance matters by establishing the animal as a coherent, embodied being. This sense of coherence will be an important aspect of the discussion of human and animal selfhood in the next chapter.

Behavior

Along with physical appearance, animals' behavior is a significant factor in decisions to adopt. As was the case with appearance, much of the social psychology that applies in our relationships with humans applies here. For instance, research consistently shows that we like people who agree with us better than those who disagree with us (see Byrne 1969; also see Aronson 1999 for a review). The more similar another person's views are to our own, the more we tend to like that person. We like people who think along the same lines because they validate our ideas. To generalize broadly, people who help us believe that we are right give us maximum social rewards at minimal risk. It is gratifying to be right, and we can increase our chances of being right by surrounding ourselves with people who share our views and beliefs.

People tend to like animals who corroborate their idea of "appropriate" dog or cat behavior.[7] This reiterates the social-psychological research on similarity and liking: We tend to like people who think or behave like us, and the same holds for animals. The reasoning follows these lines: "This cat (or dog) behaves exactly how I think cats (or dogs) should behave. Therefore, I like this animal." At the same time, behavior is an indicator of the animal as a minded being. Behavior reveals the animals' subjectivity, communicating what the animal is "really" like. This becomes crucial when behavior communicates affection for or attention to (or, more often, a combination of these) the person.

However, predicting an animal's behavior from what he or she is like while confined in a strange setting is often difficult. Although to some extent behavior in confinement indicates what an animal is generally like, animals' behavior at The Shelter is not necessarily indicative of how he or she will behave in a home. Many animals do not like confinement and may respond to it with behavior that visitors find unappealing. Dogs will often try to attract attention by barking and jumping up on the kennel, both of which turn people off. In addition, sheltered dogs must grow accustomed to being stared at by strangers, which can feel threatening from a dog's perspective. For dogs, as for humans, eye contact can have different meanings. Just as humans can send glances that say, "Don't you dare!" eye contact can be a dominance challenge to dogs. However, it is not always a challenge, and a companion dog must be willing to engage in friendly eye contact (see Derr 1997, 325). One of the easiest ways to get a dog to look at you is to say his or her name. In a shelter environment, this works only part of the time. Because many dogs arrive as strays, they do not know their names. Consequently, when people approach a kennel, read the card, and say, "Hi, Sparky," it frequently means nothing to the dog, who has had that name for only a few days. When dogs do not respond, people often assume they are uninterested. This is especially relevant given the little time that people spend interacting with the animals. In an attempt to encourage dogs to engage with clients, the canine-socialization efforts at The Shelter include work on eye contact.[8]

In the case of cats, some hide or become aggressive in confinement. To meet potential adopters, we remove the cats from their kennels and take them into a room for introductions. Many cats perform beautifully, rubbing their heads and flanks against their potential adopters, jumping into their laps, or playing with an offered toy. However, many cats who would make fine companions but need time to adjust to new environments come across as aloof or frightened. They freeze or hide. When the interaction is forced, cats can retaliate with claws and teeth, which people interpret as "She doesn't like me," instead of realizing that they (the humans) have overstepped the cat's boundaries.

Meeting a prospective companion animal in any shelter thus involves a bit of risk. Whereas it is usually the case that a dog or cat who is easygoing and calm in a shelter environment will behave the same way in a home, it does not necessarily follow that an animal who does poorly in a shelter will also do poorly at home. Alger and Alger (2003, 162) offer an example of a cat who was depressed and unresponsive in the shelter they studied but became playful and active once he had a home. Moreover, it is important to keep in mind that, in a shelter, the people also are not behaving as they would at home. Studies of sheltered dogs agree that their behavior is influenced by the behavior of the people who interact with them (Wells and Hepper 1999, 2001). Likewise, "it seems logical to assume that visitors' behavior may be equally influenced by the behavior of the dogs" (Wells and Hepper 2001, 16). In short, for animals and people, the introduction process involves two parties who are not able to disclose fully what they will be like in another setting.

Planners and Behavior

Planners usually sought specific behavior or temperament in an animal. Very often, they explicitly wanted a companion whose behavior was similar to their own. Runners wanted athletic dogs; families with active children wanted lively kittens. One planner visited the shelter once a week for about a month until a particular dog's calm kennel behavior attracted him. He explained his attraction:

> *I think of myself as being quiet and just low-key. I read a lot and I spend a lot of time at home, and I wanted a dog who was intelligent and who had it together. When I saw [the dog], he was just hanging out in his kennel, and he seemed like he was smart and mellow. I knew he was what I had been looking for.*

Fortunately, the dog's calmness was not just coincidental, for when this man and I spoke again several months later, the match had turned out to be a good one.

Very often, behavior is a factor in planners' decisions not to adopt. They reject the cat who does not play with the string they are dangling because they want a playful cat. For example, one planner rushed to meet a young dog whose picture he had seen on the website. He had wanted and waited for a dog like her, and judging by appearances, she was his type. However, in person he got a different impression. As she jumped up to greet him in typical enthusiastic puppy fashion, she scratched his bare legs. She tugged at anything she could find: the leash, the man's shoelaces, and the hem of his shorts. She barked and tried to play through the fence with the dog in the adjacent yard. By demonstrating that there was a real puppy behind the good looks, the dog showed this planner just how much training and commitment she would require. The man left without a dog that day, realizing that this one would involve more than he was prepared to undertake. In many such instances, behavior provides a good reality check for people who have little sense of what guardianship involves.

The Impartial and Behavior

For the impartial, behavior usually outweighed appearance in importance. They usually came looking for the "right" dog or cat, and what that meant often depended on the adopters' situations. For example, an impartial family looking for a dog who will play enthusiastically with children will be attracted to a different animal from the one that attracts the impartial older couple who wants a lap dog. Likewise, the couple who wants a "nice," playful kitten will draw from a pool of animals that is different from the one for the single woman whose idea of "nice" is a quiet, adult cat.

People in the subset of the impartial that I call the smitten will often overlook or accept problematic behavior. Although this can mean that some take animals home without really being prepared to cope with and address problems, it can also mean that some are simply willing to work through behavioral issues to continue living with the animal they have fallen in love with. Consequently, they gain valuable animal capital. For instance, I am one of the smitten. In May 1999, after living with cats for all of my adult life, I fell in love with a dog. I had not planned to adopt a dog at all, which was one of the reasons I did not hesitate to work with dogs as a volunteer. To my surprise, however, I not only adopted a dog, but I adopted one whose behavior was, by all accounts, challenging. To say that Skipper is high-strung is to put it mildly. Most people who know me would describe me as calm. The dog I fell in love with and adopted spun in circles in his kennel, jumped up on and nipped at people, and was, as I have affectionately described him, a "noodle head." I should also mention that he was not housebroken. His behavior could not have been further from ideal. Although he and I have both come far through training, he still has numerous behavioral issues that demand tolerance and patience. Yet we are inseparable companions; he is beside my chair as I write. I have observed the same phenomenon among other adopters. For example, one woman described this kind of attraction to her cat in better terms than I could have asked for:

> *She's an owl; I'm a lark. She's athletic, always moving, and I'm a couch potato. She's slender and elegant, and I'm a chunk. We are such complete opposites in every way that I am amazed that this creature loves me. It's just awesome.*

Along similar lines, another woman described meeting her dog:

> *When I saw him at The Shelter, we went out into one of the yards, and he was just so, I don't know, so engaged with everything, so alert. He sniffed everywhere, and then he heard something and looked, and I could tell that he was looking and listening with his whole body. If he were a person, other people would talk about his zest for life or something like that. I'm kind of, I don't know, sometimes I just don't pay attention to things around me, and when*

> *I saw [the dog], I knew that we'd be a good match for each other. I started throwing a tennis ball for him, and he was totally into it. The tennis ball and me* [sic] *were the only things in the world for him. I really wanted to be around that.*

As tempting as it is to shrug and simply say, "Opposites attract," I have tried to decipher what is behind this homespun wisdom. Exactly why do opposites attract? It contradicts the claim I made earlier that we usually like people—and animals—who are like us. However, as nice as it is to be liked by those who are like us, it is even better to be liked by those who are not like us (see Aronson 1999, 392). When someone with whom we differ likes us nevertheless, it implies that he or she sees something special in us, something worthwhile despite our differences. Perhaps the same thing occurs between people and animals. Although I have liked many calm, quiet animals, I was strongly—and strangely—flattered by how a crazy dog seemed to like me. I have heard similar reports from others who are smitten, as well. Here, for instance, a man sums up his attraction to his cat: "We're an odd couple, I knew from the start, but for some reason she likes me. How could I resist that?" His comments call attention to another important factor in attraction to animals: We like animals who like us, just as we like people who like us. Because behavior demonstrates liking, I turn now to a discussion of the role of emotions in adoption decisions.

THE "CONNECTION"

Ultimately, the most important aspect of our relationships with companion animals is neither behavior nor appearance. All adopters talked about feeling a "connection" with the animal. Planners listed it among other qualities; they would not fall head over heels in love with an animal who was not "right" for them, but once they found a dog or cat who met their criteria, they, too, spoke of a "connection" with the animal. The smitten, in contrast, felt the "connection" from the start. I put the word "connection" in quotation marks for two reasons. First, I do it to emphasize that it was the word used by the people I spoke with, and not my interpretation of their experience. So many people used the same word that I took their use of it seriously as data. In other words, if people claimed to feel a "connection," I believed them. My goal was not to find out whether they "really" connected with animals, but to find out what that experience felt like and meant for them. Because emotions are subjective, people generally take others' claims about them as truth. If I say that I love someone or that I feel happy or angry or hurt, other people have no way to verify my feelings. Like claims about other subjective experiences, such as pain, claims about emotions must be honored. They do not require any kind of verification; nor could we ever expect to develop a means by which to obtain it. Jürgen Gerhards (1989, 749) has described emotion as

> the modern *a priori*; it is the principle that does not fail when all other principles do. Emotions can claim authenticity, since any actor can personally attest that he or she has them, without others being able to refute the claim.

In this spirit, I cannot know whether the emotions someone describes are "true," in any objective sense. What matters is that they are true at that moment for the person who claims to be feeling them (see Plummer 1983, 105; Irvine 1999).

I also put the word in quotation marks to highlight how its consistent use by all types of people demonstrates the existence of an emotional vocabulary pertaining to interaction with animals. An "emotional vocabulary" is the language used to describe feelings. It is part of the larger "emotional culture," which includes the norms (known as "feeling rules" and "expression rules") that indicate appropriate and desirable feelings (and expression thereof) for given situations (see Goffman 1959; Hochschild 1975, 1983; Gordon 1981; and Stearns 1989a, 1989b, 1994). The contemporary American emotional vocabulary includes rather traditional and straightforward words, such as "love" and "anger," but it also includes newer terms, such as "freaked out," "stressed," and "blown away." As new emotional states emerge within a culture, new vocabularies arise to describe them. "Road rage" is an example. The notion of a "connection" with animals is likewise a reflection of a particular time and emotional culture. As I explained in Chapter 2, feeling emotionally close to animals was not a desirable state for much of Western history. Indeed, with some exceptions, people often paid for such feelings with their lives. A "connection" to animals was sociologically and psychologically unlikely until relatively recently.

The Signal Function of Emotions

When potential adopters say that they knew they had found the right animal because they felt a "connection," they are referring to the capacity of emotions to serve as signals. The signal function, first outlined by Freud, refers to the way in which feelings help shape our perceptions of the world around us, as when the knot in our stomach signals nervousness (see also Hochschild 1983). However, we feel nervous, or happy, or angry only in the context of interaction. The idea of having raw, uninterpreted emotion outside of a social context makes no sense. Emotions function as signals in conjunction with the expectations we have about situations. The signal "involves a juxtaposition of what we see with what we expect to see" (Hochschild 1983, 221). Thus, we feel anxious before giving a presentation or happy to be with our loved ones, for example, because we hold those expectations for those situations. The emotion helps define situations as tense or enjoyable, but we also use other information along the way.

For adopters, the expectation that made sense of the emotional signal had to do with the potential for a bond with an animal. After all, they went to The Shelter hoping to adopt a dog or a cat, and presumably one with whom they could have a "connection." Certain animals communicated that potential by indicating that they liked the person. This is why an animal with all the right characteristics sometimes was not "right" for an adopter: the animal could not have cared less about him or her. Thus, the necessary factor in people's decisions to adopt was whether they thought the animal liked them. The affection signaled a "connection"—or, at least, the potential for one—which sealed the match.

How do animals communicate liking? They use their bodies. Dogs and cats have ways to make it very clear when they want to be with us. Cats use flank and head rubbing, especially as a greeting. This sends a message of affection, as does purring, which communicates pleasure in another's presence. Dogs stay close, wag their tails, and relax their faces and ears. Animals also show interest, if not affection, through attentiveness. This can often involve physical closeness and contact. People who discover the place in which a dog or cat particularly likes to be scratched or rubbed are often flattered by the reaction. The experience of "hitting the spot" can imply that the person somehow knows the animal.

Planners and Connection

Planners wanted a connection in addition to the other characteristics they sought in a companion animal. The other characteristics by themselves were not sufficient indicators of the "right" dog or cat. For example, here a planner describes meeting his dog:

> *There were two dogs there that day that I was interested in, both Labs, and both of them looked good, you know, on paper. I've had a couple of Labs before and I love them. Great family dogs. I met both these dogs, and they were both great dogs and I suppose I would have adopted the first one. He was a bit younger and better looking, but we had no connection. We went outside together, and he pretty much ignored me. So at first I thought, well, he's a nice dog; I'll just have to give it time. But then I met the other dog, and we really connected right away.*

In a sense, the relationship between some planners and their companion animals reminded me of arranged marriages. They had requirements for compatibility that did not include head-over-heels love at first sight. Indeed, they thought love at first sight was a poor basis for a match.

They expected to feel *something* (a "connection") for the animal, but that feeling alone was not enough. For example, one planner described a "connection" to a cat who was nevertheless not right for her:

> *I look at the website all the time, and I came in a bunch of times when you had Siamese cats because I've been wanting one. I met a couple of them over time. I remember this one, she was just so sweet. I really felt like we had a connection. But she was too old and she had long hair. Finally, I met [another cat], and that was it. She had the look I wanted. She was the right age, the right size. There was also this connection between us. It was right from the start. There was just something there.*

Planners' caution about using emotions as a basis for adopting an animal reveals something important about American middle-class emotional culture. In particular, it reveals ambivalence about the reliability of emotions as guides to action. Before I take up that discussion, let us look at the experience of the impartial for comparison.

The Impartial and Connection

The impartial placed more weight on the "connection" and its signal function than planners did. They read the potential for "connection" just as planners did, through the animals' willingness to make contact. For example, one client sought a cat to keep her company after her husband died. She looked at all the kennels and said she could easily have taken any one of them, but the cat she adopted was the one who put her tiny paw out through the bars as if to introduce herself. Among both dogs and cats, a willingness to engage with people is an important way to communicate "I like you."

Because the impartial arrive at The Shelter with few or no specific physical preferences for their animals, the "connection" is essential. Often, the feeling can be quite clear and dramatic. One of my favorite examples is Marc Bekoff's story of what he felt when he met and adopted his companion dog, Jethro. Bekoff has pioneered the scientific investigation of the cognitive and emotional lives of animals. His story of unmistakable bonding with a sheltered dog defies scientific methods:

> I simply walked into the Humane Society and felt this instantaneous connection. It was as though my whole body was magnetized to the very back corner of the room, where I found this dog staring up at me. I can't explain it, but it was a warm, wonderful feeling that completely overcame me, something I felt seventy-five feet across all those cages. I knew I'd found a companion even before I saw him. I'm a trained scientist. I'm not supposed to have these kind of universal connections—they put me in hot water with my associates. But I do. (As quoted in Schoen 2001, 172)

Like Bekoff, others described feeling a "spark" or a "pull" from a particular animal. They also regularly attributed this to being destined to be with an animal, of knowing immediately that this was the dog or cat for them.

In some cases, the feeling of "connection" not only communicated "I like you"; it also sent a message of "I am like you," demonstrating the link between similarity and liking. For instance, the woman in the third excerpt at the start of this chapter felt undeniably drawn to a cat who, by all accounts, was not very attractive. He had been a stray. We knew little about him except that he had arrived at The Shelter through the kindness of a man who worried about the condition of the cat he had seen hanging around his store. We did not know the cat's age, but he was no youngster—probably as old as fifteen. He had fought with other cats before arriving at The Shelter, and the vets had drained an abscess and stitched up a gash on his head. This had required shaving an area down to the skin, which revealed the scars of previous fights that had taken place over years and years. The cat had been recuperating in foster care, so his fur had begun to grow back, but he still looked like the veteran of the streets that he was. Moreover, he had only a few teeth left. Nevertheless, he found a home that day. The woman who felt a "connection" to him had suffered a head injury and had consequently had to relearn how to talk and do other things most of us take for granted.

Two surgeries had left scars on her head. She felt as though she recognized a soul mate that day. For her, as for many others considering animals, that feeling meant, "I'm the one. Take me! We're *meant* to be together." In the same vein, Alger and Alger (2003, 159) found that the most common reason people adopted particular cats was the belief that the cat had chosen them.

AMBIVALENCE IN THE EMOTIONAL CULTURE

Planners and the impartial illustrate mixed messages conveyed by the emotional culture. In many arenas of life, we rely on emotions to tell us what is "really" going on "in our guts." We believe that feelings reveal our authentic desires and intentions. Hochschild (1983, 190) puts it well in saying that, "as a culture, we have begun to place an unprecedented value on spontaneous, 'natural' feeling." This represents a paradigm known as the "organismic" model of emotions, generated by the work of Darwin, Freud, and James.[9] It equates emotions with instinct, capable of prompting action consistent with an allegedly "true" self. From the organismic perspective, emotion exists apart from both the "feeling rules" that pertain to the situation and the introspection of the person experiencing the emotion. Consequently, emotion constitutes an exceptionally reliable guide to interaction. If emotions indicate the "true" self, then in the absence of any apparent "contamination" from outside factors, the experience is undeniably powerful.

However, alongside this respect for "natural" emotions lies suspicion. We may respect emotions as instinctual, but we simultaneously label some as dangerous, especially if they are intense, because they jeopardize rationality. For example, most people would probably consider an emotion such as anger dangerous and would agree that it should never serve as the basis for action. Using the current emotional vocabulary, the best thing to do with anger is to "manage" it. Likewise, most people would consider guilt a poor guide to sound behavior, because it is thought to be both cause and consequence of psychological damage (see Irvine 1999). This raises questions about which emotions are indicators of the "true" self, and under what circumstances. Historians have analyzed this by tracing the rise and fall of specific emotions within the American emotional culture. Over the course of the twentieth century, the American middle class gradually adopted a more moderate emotional style that signifies "a uniformly cool, controlled personality" (Stearns 1994, 263).[10] The clear but implicit preference is for a display that suggests that one has everything under control. Studies have documented similar shifts toward emotional restraint and the sanctioning of intensity in several industrialized Western nations (see de Swaan 1981; Gerhards 1989; Wouters 1991), but the United States leads overall (see Sommers 1984).[11] In short, the cautious emotional culture tempers our urge to act purely or primarily on the basis of feelings

To return to the analysis of the adopters, planners went looking for a particular type of animal, and the "connection" sealed the bond. The feeling of "connection" was important, but only in the presence of other characteristics of the animal. In other words, the "connection" was necessary but not sufficient to make a match. In contrast, the impartial had no specific type of animal in mind. The emotional experience thus came in the apparent absence of expectations, like pure, unadulterated, "from the gut" intuition.

CONNECTION AND MOTIVE

The experience of feeling destined to adopt a particular animal has a powerful function. I came to see it as an avowal of motive, in terms set out by Max Weber (1968 [1922]) and later elaborated by C. Wright Mills. Motive refers to "a complex of meaning, which appears to the actor himself or the observer to be an adequate ground for his conduct" (Mills 1940, 906). Motives are answers to "how" and "why" questions, but they are not merely descriptions of or reasons for our actions. They are ways of influencing others and oneself to see our conduct in a particular light. We construct motives when a question might arise about what we have done: "Why did you adopt *this* dog?" for example, or "I hadn't realized you liked Siamese cats." "How did you decide on that one?" In the face of such questions, or even the anticipation of them, "I was meant to" is an indisputable motive. Moreover, it is consistent with the organismic model of emotions. A planner, for example, would be unlikely to talk about a magnetic pull to an animal apart from the characteristics that made the animal the "right" choice, but someone who is impartial must nevertheless find a way to account for his or her relatively random selection. In addition, if I believe that an animal was somehow "meant" to be with me, then a promising implication of holding this belief is that the bond will last and may be especially meaningful to me. Under the best circumstances, this feeling will take the relationship through the hard times that any relationship brings with it. For example, my belief that Skipper and I were meant for each other has enabled me to accept many behaviors that someone else might not. My motives have changed over time, which supports Mills's claim that they are strategies of action rather than literal reasons. Initially, I would have explained my attraction to Skipper as "I was supposed to be his person." Now, three years later, in the midst of this writing, I have caught myself saying that that Skipper and I got together because I needed a teacher. Had I adopted a different dog, I would not have learned as much about canine behavior, or about my own behavior in relationship to a dog.

Ultimately, all adopters seek to "connect" with an animal, although they go about it differently. Regardless of whether they sought a particular type of animal or fell in love unexpectedly, the common denominator was feeling that they and the animal were "meant for each other." It is easy to dismiss this as anthropomorphic projection, but that does the experience a disservice. For if the feeling of "connection" with an animal came solely through anthropomorphizing, then people could project almost anything onto an animal. Adopting an animal would simply involve finding a cat of the right color or a dog who could obey the right commands. Although there are surely adoptions that occur in this way, much more is usually involved in making a good match. Often, the cat or dog who seems right is somehow all wrong for a particular person.

What all prospective adopters are trying to do is learn what the dog or cat is like. In other words, they are trying to get a preview of the animal's sense of self. Finding a companion dog or cat is much like the process of finding a human friend (although dogs and cats are admittedly easier to approach). Most of us have known people with whom we felt instantly familiar. You feel like you can "be yourself." More accurately, you feel like you can be the *most* yourself that you can be. You are not just projecting—or, if you are, the relationship does not last long or go smoothly. Granted, the feeling stems partly from having things in common, but it is not solely that, for if it were, most of us would probably have far more friends. What it both depends on and indicates is the "fit" between the selves of those involved. This takes place with animals, too, and the next chapter examines how animals' selves become present to us.

6

Rethinking the Self: Mead's Myopia

What are animals really like? How far can we trust our own unthinking recognition of their fear, fidelity or cleverness? How far should we accept the impulse to decree a strict division between us and them? Here then is the issue. How shall we decide?

—Stephen R. L. Clark (1982, 2)

The observations I made in the adoption areas may ring true for many readers, who perhaps identify with the adopters' feelings of connection or the visitors' pleasure in seeing and interacting with animals. Familiar as these observations may seem, however, they alone do not offer a theory for understanding our relationships with animals. The theory emerges by searching for the theme that underlies the pleasures and concerns of human–animal interaction. That theme, I argue, is the sense of self. Animals confirm and enrich that sense in us. To do so, animals must also have selves. In other words, animals help create human identity by interacting with us as selves in relation to us.

Let me recap some examples for the sake of making my point. The interaction that visitors and adopters had with animals was not limited to looking at them; it involved talking to them and about them, as well. Everyone—regardless of intention to adopt—wanted to know their names and their stories and, whenever possible, to hold, touch, and play with them outside their kennels. I mentioned that some even came regularly and developed relationships with long-term Shelter residents. People who could not have animals of their own nevertheless came to visit The Shelter, suggesting that interaction with animals can become so vital that going without becomes unthinkable. Moreover, many people who had browsed The Shelter's website wanted to meet the animals whom they had only seen in thumbnail photos. In other words, people did not simply want to know that the animals existed; they felt drawn to interact with them directly. The structure of people's interaction in the adoption areas indicates that they include animals among the others who are available for relationships.

Visitors and adopters were concerned about learning the animals' histories and, in the case of adoption, to provide more of what they needed. Concern for the well-being of the animals also manifested itself in people's words, such as "I feel so sorry for them," and "I wish I could take them all home." People commonly expressed concern for the animals' freedom, as in "I hate to see them in cages." In addition, people were genuinely concerned when animals seemed afraid or were obviously recovering from an operation or injury. For example, The Shelter's male dogs must routinely wear E-collars (or Elizabethan collars) to prevent them from licking at and potentially disturbing their sutures after they are neutered. They wear these for only a few days following surgery, but people consistently ask what happened to the "poor dog" with the white cone around his head. The analogous situation for cats is when stray or neglected animals come in with fur so badly matted that they require shaving, sometimes leaving fur only on the head, paws, and tail. Shelter staff and volunteers grow so accustomed to answering questions about the "poor kitty" that we often add notes explaining the "bad hair day" on the cats' kennel cards. Thus, visitors and adopters alike expressed interest in the animals' needs and well-being.

Finally, interaction in the adoption areas also revealed that people who had had animals previously, some for most of their lives, were less susceptible to the "cute response" and interacted with a wider range of animals. In other words, animal capital paid a high rate of return. Moreover, as I have mentioned, the animals served as social facilitators, sparking conversations among visitors and thus encouraging the use of interactional skills with other people. Consequently, relationships with animals may actually increase our interactional abilities.

Overall, the structure of interaction in the adoption areas reveals three main components: Seeking relationships with (animal) others; expressing concern for animals' well-being; and engaging in increasingly complex interaction. The theme that underlies these components is the self. More specifically, they are behaviors or activities that manifest goals of the self (see Myers 1998). The notion that the self has goals emphasizes that “”having” or “being” a self allows—or even requires—us to do things. One of the things that the self allows us to do is to “go on being” (Winnicott 1958). In this sense, then, a primary goal of the self is continuity. If the self does not survive, if it has no continuity and simply disappears, then there is nothing more to discuss. Assuming, then, that much of what we do, we do to go on being, we can pursue the goal of continuity in many different ways. One of these involves forming relationships. We know from Mead and others that the self emerges through relationships. Granted, once the self has developed, it can exist without relationships, so that even the person in solitary confinement continues to have a sense of self. However, relationships allow us to develop a mutual history that is simultaneously a history of the self (see Irvine 1999). Therefore, concern for the well-being of others, expressed through an interest in their needs, ensures the continuity that provides the relationships on which the self depends. As Myers (1998, 50) explains, “Having a self predicts interest in significant others.”

If relationships are essential for the self, it would also be important to increase the skills that make relationships possible. After all, maintaining relationships requires the use of the interactional skills that foster relationships in the first place. One of the signs of a good relationship is that the improvement of interactional skills then helps us in additional relationships (see Csikszentmihalyi 1990). Simply maintaining relationships at status quo is better than nothing, but still not enough. The quality of the interaction is also crucial. As Mihaly Csikszentmihalyi (1997, 43) puts it, “Even the passive superficial conversations at a neighborhood bar can stave off depression. But for real growth, it is necessary to find people whose opinions are interesting and whose conversation is stimulating." Good relationships happen when both people have an investment in the interaction, when they have engaging conversations, when they ask questions and take interest in the other. Good relationships stretch our interactional abilities by requiring us to see things in new ways and remain open to surprises. They offer "new information—incongruities, interruptions of expectations, challenges—in the context of familiar otherness" (Myers 1998, 78). The payoff is usually immediate. Both people come away from the interaction with a sense of where they stand in the relationship and with the feeling that it is worth the investment it takes to maintain it. These words do not really do the experience justice, however, for there is much about good relationships that is pre-linguistic and even pre-cognitive. Think about the last time you really "clicked" with someone. Now try to put the reasons you liked that person into words. Although there are probably elements of the interaction that you can easily describe, much of it exists at the level of pure feeling. Certain relationships click because they challenge our interactional skills just enough and consequently increase our ability to *have* relationships. As is the case with physical exercise, we build "muscle" that equips us for further challenges. Eventually, the exercise itself becomes intrinsically rewarding. This process of interactional challenge and reward takes place in good relationships, whether with other people or with animals.

The structure of interaction between people and animals (seeking relationships with animals, demonstrating concern for their well-being, and engaging in increasingly complex interaction) revealed that animals mean something for the experience of selfhood. The question that arises has to do with *how* they "mean something." Moreover, how do animals differ from the other things that also contribute to our sense of self?

They key is the subjective presence of the other. The interaction must seem to have a source, and we must see the other as having a mind, beliefs, and desires, just as we do. This not only confirms the other's sense of self to us; it also confirms our own. How do we sense an other's subjective presence? With people, we can rely on self-reports. However, these reveal more about the norms of self-reports than about anything else. The answers reveal the influence of what people know to be good, desirable, acceptable depictions of the self. They reveal a self digested in consciousness and shaped by language. They indicate how people talk and think about the self. A stronger objection is that, even with other people, we simply do not rely on language first or foremost for information about selfhood. As Goffman (1959) wrote, only part of the self is conveyed through "impressions given"; others appear through "impressions given off." Nevertheless, the idea that language is necessary for subjectivity is deeply entrenched in the research on the self.

GEORGE, MEET WASHOE, KOKO, AND ALEX

George Herbert Mead (1962 [1934]) is largely responsible for giving language the prominent role it has in social psychology. Language has constituted the "official" barrier between humans and non-humans because, among other things, it enables us to understand and communicate the symbols for self, such as our names and the names of other objects. It allows us to talk about ourselves, picture ourselves, remind ourselves, and we even speak of kicking ourselves or patting ourselves on the back. Although language certainly allows us to do things that other species cannot, as we shall see, Mead (and others after him) "gave language a role it cannot fulfill" (Myers 1998, 121; see also Hanson 1986).

Mead advanced a version of the rationalist tradition that started with Aristotle and was endorsed by Descartes, whose claim, "I think; therefore, I am," required the ability to talk about thinking.[1] As Myers (1998, 39) puts it, "Mead explicitly replaced the Aristotelian and Cartesian markers of human difference—'soul' or 'mind' with a secularized version—language behavior." Although Mead acknowledged that animals have their own social arrangements, he claimed that their interaction takes the form of a "conversation of gestures." The term refers to primitive, instinctual acts, such as when a dog growls at another dog or a cat hisses at a rival. Mead considered the "conversation of gestures" insignificant because there is allegedly only one possible response to the growl. In Mead's view, the dog who growls or the cat who hisses is simply sending out a stimulus that produces a "Back off!" response in all other dogs and cats. As John Hewitt (2000, 9) explains, "In no sense does either [animal] 'decide' or 'make up its mind' to act in a certain way." In this perspective, the behavior of animals may be goal-directed in that it aims at getting food or a mate, or defending territory, but it supposedly lacks the premeditation and shared meaning that characterize human behavior.[2]

In contrast, humans use the vocal "gestures" we know as language. Mead referred to language as "significant symbols": *significant* because they purportedly produce a common definition of a situation in both sender and receiver. In doing so, language allows people to anticipate the consequences of their actions, consider alternatives, and coordinate their actions with others. The example most often used is when someone shouts "Fire!" According to Mead, the word does not simply produce a flight response, as he thought the growl did for the dog. Instead, it produces an image of the situation and one's position in it, which then allows the individual to consider several possible plans of action—or to decide not to act but to wait and see what others are doing. Through this process of imagining and controlling responses, we constitute the self as an object.

In Mead's formulation, then, spoken language allows us to develop selves by seeing ourselves as objects and giving meaning to situations. Moreover, Mead maintained that the capacity for language was both cause and consequence of the human mind. We are born with the ability to use language, and in learning to do so, we imagine different perspectives and potential plans of action. In addition, as we imagine situations, control our behavior, and coordinate our actions with others, we participate in and recreate society—hence, the title of Mead's key work, *Mind, Self and Society*. Lacking the capacity to use significant symbols, Mead renders animals incapable of having any meaningful social behavior. As Mead put it, "The animal has no mind, no thought, and hence there is no meaning [in animal behavior] in the significant or self-conscious sense" (as quoted in Strauss 1964, 168). Any "mindedness" we humans ascribe to animals is, in Mead's view, simply anthropomorphic projection.

The pitfalls in this view are numerous. First, in calling spoken language "significant" and everything else "merely" instinctual, Mead (and, consequently, social psychology) reproduced the Cartesian error. Mead established two states of consciousness: one for those who could converse about it, and another, lesser form for those who could not. Both Mead and Descartes had reasons for holding this belief, but whatever those reasons were, they had no empirical basis. Consciously or otherwise, both were staking an ideological claim that safeguarded human superiority. Descartes' views justified the exploitation of animals prevalent in his day, and Mead's ideas helped an insecure social science distinguish its human subject matter from "mere" animals, who had behavior, but not meaningful interaction. However convenient it was to use spoken language as a dividing line, doing so represents a definitional move, not an empirical one (see Allen and Bekoff 1997).

Second, since Mead's day, copious research has debunked the idea that spoken language is a uniquely human capacity. Numerous species have learned to communicate using symbols, computer keyboards, and American Sign Language (ASL). Washoe, the first chimpanzee to learn sign language, had a working vocabulary of 140 ASL gestures and knew twice as many two-sign combinations. Koko, the lowland gorilla raised by the psychologist Penny Patterson, could use more than 600 signs. Critics say that it is one thing to learn the names for things and quite another to understand the meaning of language and use it to innovate on one's own. However, Koko and other primates have demonstrated that they understand and use sign language in contexts other than the basic drills in which they name objects shown them by their humans. For example, *The Education of Koko* (Patterson and Linden 1981) describes an instance observed by Barbara Hiller, who worked with Koko for many years. One day, Hiller saw Koko playing with several white towels and making the ASL gesture for the color red. Seeing the white towels, Hiller corrected Koko, who only signed "red" larger, which is the means of exaggerating in sign language. Hiller again corrected Koko, who once more signed "red" larger and more emphatically—and then plucked some red lint off the white towel for Hiller to see.

The capacity for symbol use hardly comes as a surprise in the case of primates, who are the "border species" between humans and animals. However, other species, too, have shown themselves capable of using the tool that long stood for human superiority. Pepperberg's (1991) long-term research with Alex, a twenty-two-year-old African Grey parrot, reveals that birds can use language. Alex has also shown that his abilities go far beyond simply naming objects, which he must do in endless drills as part of Pepperberg's research. Alex has shown himself capable of violating rules, which indicates that he not only knows the rules, but also understands the abstract and complex idea of distorting them. For instance, one trick Alex uses to express his irritation with the language drills is to give incorrect answers so many times that it becomes obvious that he is doing so on purpose. Moreover, he will give different incorrect answers each time, so that the exasperated researchers eventually give up. Although this could be statistical chance rather than an act of defiance, the odds of giving so many wrong answers in a row are slim (see Linden 1999, 40).

Third, the emphasis on spoken language neglects the importance of other forms of communication. Although we cannot ask animals—or mute humans, for that matter—to tell us about themselves, we gather abundant evidence from other behaviors, such as the structure of interaction, as I have proposed. Scholars have already done so in the case of the mentally disabled (e.g., Pollner and McDonald-Wikler 1985; Bogdan and Taylor 1989), Alzheimer's patients (e.g., Gubrium 1986), and infants (e.g., Brazelton 1984; Stern 1985). These studies examine how caregivers articulate the minds and selves of those who have no capacity for self-expression. They show how "others literally 'do' the minds and selves" of those who cannot speak, building on an emerging sense of attitudes (in the case of infants and the retarded) or sustaining an established sense of identity (in the case of Alzheimer's patients) (Holstein and Gubrium 2000, 152). Even among other people who can express themselves, we understand how much a look, a tensing of the shoulders, a wink, or a sigh can mean. Many of us learned this difference early; I surely am not alone in having had a mother who told me that it was not what I said but how I said it that got me into trouble. If Mead was right, then scores of us would have been free to sass our parents as we pleased, for they would have understood what we really meant. But, as Myers (1998, 121) explains,

> [Mead] said it is the simple audibility of the spoken word that provides the person a way to identify with or have access to the perceptions of others. There are problems here, for actually we do not hear our own words exactly as others do—due to anatomy, physics, and differing perspective on what is meant. Words have to be interpreted; they do not work automatically . . . with words, we can express ourselves or something characteristic of the self can be summed up—but if this is how the self becomes available then it is in a different respect than mere audition that language makes the self reflexively available.

Clearly, the overemphasis on language eliminates a considerable amount of interaction as a source of information that contributes to selfhood. Moreover, it restricts the significant interactants to other people. If we can at least tentatively agree that factors beyond spoken language matter, then animals can participate in the creation of human selfhood. However, this still leaves the two questions I raised earlier unanswered. I take up the question of how animals participate in the creation of the self in the following chapters. The second question, which concerns how animals differ from the other things that contribute to selfhood, I will tackle here.

Thus far, numerous other things can—and do—meet the goals of the self that I sketched out. Art, music, hobbies, nature, and books "participate in the creation of human selfhood." I argue that there is something different about interaction with animate others. Research confirms that this is indeed the case. For example, the behaviorist James Watson demonstrated conditioning among children in a famous experiment on an eleven-month-old boy known in the literature as "Little Albert." Watson paired a loud noise with the sight of a live rat and consequently conditioned the child to fear the rat. According to behaviorism, it should not matter whether the object was animate or inanimate; the point was that the response could be conditioned. The psychologist Elsie Bregman later showed that this assumption was flawed. She replicated Watson's experiment using blocks of wood and rags but found that the initial fear would diminish, whereas that was not the case when a rat had evoked it. Thus, even infants can make a distinction between animate and inanimate.[3] The lesson in this is that animals offer interaction that has something in common with human interaction (although they offer a great deal that is different, as well). In other words, they offer something that blocks and rags cannot. I saw evidence of this among children in my research. In the Mobile Adoption Unit, a hollow, molded-plastic dog-and-cat "statue" about three feet high serves as a bank to collect donations. The bank fascinates young children—at first. They touch the dog's nose and ears and "meow" at the cat. Sometimes the dog wears a bandanna or a Gentle Leader (head halter), and they try to remove or adjust them. They pet both animals and poke their fingers in the hole in the dog's head that allows people to insert money. However, their eventual preference for the animate, or the "real" animals, to the inanimate statue becomes clear. Although they often spend several minutes investigating and talking to the plastic dog and cat, they spend as much time with the real ones as mom and dad will allow.

The difference between how animals "mean something" for human selfhood and how inanimate objects do so has to do with the subjective presence of the other. The interaction must seem to have a source, and we must see the other as having a mind, intentions, and desires, just as we do. This not only confirms the other's sense of self to us; it also confirms our own. How do we sense the subjective presence of an other? With other people, we can rely on self-reports because we share language, but, as mentioned earlier, these reveal more about the norms of self-reports than anything else. For example, if I ask you to describe yourself, your answers will show me a self digested in consciousness and shaped by language. Although this is interesting in and of itself, it is knowledge once removed. It indicates how people talk and think about the self. It reveals nothing about subjective experience. Thus, even with other people, we cannot observe subjectivity directly. We have no direct access to it. We perceive it *indirectly* in the course of interaction. As we will see in the next chapter, this includes our interaction with animals.

7

Self versus Other: The Core Self

Animals are such agreeable friends. They ask no questions, they pass no criticisms.

—George Eliot, *Scenes of Clerical Life* (1880, 138)

We know the selves of animals in the same ways that we know the selves of other people. Two processes occur simultaneously. First, animals' subjectivity becomes available to us because the elements of a core self become visible through interaction. Animals, like people, are born with the capacities for core selfhood. These capacities then allow for and depend on interaction and relationships, which in turn elucidate and engage additional senses of self. The capacities for the core self do not depend on language. In humans, these core capacities appear in the first few months of life (see Stern 1985). Human development takes us into a stage of language acquisition that adds verbal dimensions to these core elements, but the core self is pre-verbal. The case can thus be made for its presence among animals, who have the same physiology of the brain, nervous system, musculature, and memory that enable the formation of a core self among humans. The self is therefore the product of both nature and nurture. The social, or "nurture," side of the equation is the traditional terrain of sociology, but there is also a well-documented "nature" side, consisting of neurobiological links between physiological characteristics and selfhood (see Damasio 1999).

A second occurrence that allows us to know the selves of animals is the confirmation of our own sense of subjectivity during interaction. Interacting with others, whether people or animals, verifies our sense of who we are, and even that we are. In Chapter 6, I described the self as having goals. I also argued that seeking relationships, engaging in increasingly complex interaction, and expressing concern for the well-being of others suggest that one of the self's goals is to enable us to "go on being." One way to see this is that the self has a functional side that allows us to do things. But selfhood also allows us to feel and to know. Thus, along with being a system of goals, we can also describe the self as a system of experiences. The two senses of self coexist and inform each other. Infants attain basic experiences of self through interaction with others. This interaction then allows us to make critical distinctions between self and other. The self–other distinction, in turn, both enables and requires further interaction and relationships.

The idea of the self as a system of experiences is adapted from William James's attempts to gain access to the subjective sense of self, or the "I." Along the way, James distinguished four features of the "I" that underlie and make us aware of it. Others have since refined these to study infant subjectivity, in particular (see Stern 1985; Myers 1998). The structure of interaction between infants and caregivers reveals these basic features even in the first months of life. Because infants cannot describe self and other, these features of subjectivity make a good departure point for the study of selfhood among animals. The four features are:

1. A sense of agency, meaning that you are the author of your actions and movements and not the author of the actions and movements of others;
2. A sense of coherence, meaning that you understand yourself as a physical whole that is the locus of agency;
3. A sense of affectivity, meaning patterned qualities of feelings that are associated with other experiences of the self;

4. A sense of self-history (or continuity), meaning that you remain to some degree the same, even while you change.

Combined, these four senses compose a "core" self, in that they are considered necessary for normal psychological functioning (see Stern 1985, 71). The four capacities not only underlie our own sense of subjective experience; they also form the basis for distinguishing subjective others. The absence of one of the four manifests itself in psychosis and other pathologies. Granted, there are additional senses of self, many of which require the acquisition of language, but it is reasonable to consider these four as essential. Let us look at them systematically.

AGENCY

The word "agency" is used in sociology in many different ways, so I want to specify how I use it.[1] Here, I refer to the capacity for self-willed action. Agency has several features, which include, in addition to initiating my own action, having some control over my own actions (I can sit when I decide to, and if you push me into a chair, that is something altogether different) and having some awareness of felt consequence (my intention to sit down in a chair brings the felt consequence of sitting). The kind of agency I refer to here implies a level of consciousness, in that an agentic being has desires, wishes, and intentions along with a sense of having those things. In other words, it is the actor's awareness of having desires or wishes that is an element of selfhood, not simply having them.[2]

Agency manifests a sense of volition or will because of a brain–body connection known as a "motor plan." In brief, this mental registration emerges before our muscle groups can carry out any voluntary motor skills that are higher than a reflex.[3] Every time we reach for a pen or turn a page, a motor plan is involved. The motor plan occurs outside of consciousness, but it can easily come into consciousness when it fails. When we intend to sit down but position ourselves incorrectly and fall, or when we want to scratch an eyebrow but poke ourselves in the eye, we know that something went wrong. The existence of the motor plan facilitates the sense of will. This, in turn, creates the sense that our actions are "ours," even though most of the time we are blissfully free of the need to think in such terms. Because our bodies and animals' bodies operate similarly, we could thus expect that animals also have a sense that their actions "belong" to them. If you have ever seen a dog or a cat attempt a jump and fall, or lose footing on a slippery floor, you have witnessed the fleeting moment of "recovery" and even the embarrassment they seem to feel, similar perhaps to what we feel when we miss a chair or poke ourselves in the eye.

Various indicators of a sense of agency appear in human infants during the first months of life (see Stern 1985). Examples that appear in this early stage include reaching for objects and hand-to-mouth skills. At around four months, infants learn to use visual information to shape the fingers to accommodate objects of particular sizes. Because agency does not depend on verbal ability, it is therefore feasible among other species. Let me offer some examples of how it becomes apparent among dogs and cats.

Strong evidence of animals' agency comes from the arena of dog training, even at the most basic level. The main thing that trainers teach dogs is to exercise self-control—and they use this very phrase (see also Sanders 1999, 138). This implies that the dog must have a sense that he or she can initiate action, because to control one's self one must first have a sense of will or volition. For animals, as in humans, agency occurs developmentally. Puppies are no more able to learn how to "sit" than infants are to walk on their own and feed themselves. The capacity to do so is nevertheless present, and training involves shaping the dog's agency in ways that will allow him or her to live among humans. Dogs have to learn to control the urge to do the first things that come into their heads, because these "first things" are usually things that humans dislike, such as jumping and barking.

At The Shelter, a considerable amount of time and effort goes into socializing dogs. On numerous occasions, I have worked with staff members to help make an under-socialized dog more adoptable. One case involved a mixed-breed puppy who could melt hearts with his comical one-ear-up-and-one-down appearance. His behavior, however, made most people think twice about adopting him. He jumped up on the gate of his kennel and barked frantically whenever anyone came into the adoption area. Our goal was to take advantage of the dog's agency and shape his behavior into that which would make him more appealing to adopters. The key to doing this was to change his understanding of the cause of the rewards he receives. Changing his understanding in this way highlights the non-verbal capacity to distinguish self from other.

A dog who jumps up on his kennel and barks receives two kinds of rewards. To the extent that the behavior is self-directed (i.e., aimed at releasing energy), it is continually rewarding. However, to the extent that it is directed at others, as an attempt to gain their attention, it is randomly rewarding. As long as the reward of attention depends on others, it will be unpredictable. Many people will avoid such a dog—as was the case with the dog in this example. The ability to distinguish between different reinforcement schedules helps us discriminate between self and other. Agency also allows for differentiation between self and other. For example, if I reach around to scratch an itch on my back, my act of scratching has felt consequences. Chief among these is relief from the itch, but another consequence is the sensation I have through the fingers that do the scratching. They feel the skin in the location of the itch, and I know that I am scratching myself. In contrast, if I ask my husband to scratch my back, the itch is relieved but I have no sense of doing the scratching. In addition, when I do the scratching, it is certain that I will hit the target. However, if my husband scratches my back, we will go through a few rounds of "Up. Down. Now to the right just a little. Now left," and even so, there is a chance that I will end up scratching the itch myself. In short, when an action is self-initiated and directed at the self, as is my scratching of my own itch, it has a constant, high reinforcement schedule. Other-initiated actions directed at the self are less predictably reinforced. The different reinforcement schedules help us distinguish between self and other.

Given that we develop the ability to distinguish between differing schedules of reinforcement as early as three months of age, it is likely that highly social animals can do so, as well, with the same implications for discriminating between self and other. To return to the example of the dog, the staff members and I had to make him aware of this distinction and show him how to increase the probability of rewards from others. To do this, we moved him out of the adoption area for a few days to reduce the foot traffic past his kennel. We scheduled regular exercise to reduce his need for the jumping. Most important, we stopped reinforcing his bad behavior. We paid attention to him only when he was quiet and had all four paws on the floor. If he barked or jumped, we moved away from his kennel. Because he had released some of his pent-up energy through exercise, the reward of attention quickly became a higher priority. Moreover, because attention depended on others, he soon learned to control himself to get it.

To be sure, the dog did not become a paragon of obedience through this brief behavior-modification exercise. However, he did begin to show signs that he could control himself, which is often all a potential adopter needs to see. One day, as a staff member and I discussed our work with this dog, she made the point clear to me when she said, "We have to get him to be able to show people that he'll be worth it." It occurred to me that, although our explicit task was to help him learn basic canine manners, our larger, albeit implicit, goal was to enable him to demonstrate to people that he had something underneath, or other than, the problem behaviors. In other words, we had to help him develop the control over his own behavior that would show people that he had—or was—a self.

Explicit training occurs less often among cats than among dogs, but it frequently occurs indirectly.[4] Anyone whose cat comes running to the sound of the can opener knows this, although there is some question of who is training whom. As Alger and Alger (2003, 21) point out, because life with cats generally involves fewer rules than does life with dogs, training is less important. Interaction between cats and guardians is highly social rather than the goal-oriented interaction between dogs and guardians. Thus, cat guardians most often talked about other evidence of feline agency. One example guardians repeatedly mentioned had to do with eye contact. "He'll just all of a sudden look at me," one man told me. "I have no idea why. But he looks over like he's just checking in." For some people, the look from a cat was what sealed the adoption decision. Several said, "I liked her (or him) because she looked at me" (see also Alger and Alger 2003, 161). Having a cat come to them also confirmed agency. "When I sit and watch TV," one guardian told me, "he'll come from wherever he is, just come striding over, and sit with me. As I watch him walk over, it's clear that he knows what he's doing."

Touch is another way cat guardians experienced their companions' agency. For instance, one guardian offered this story:

> *I went to lie down on the couch for a nap, and I had my head propped up on the pillow at one end and my arm was up like this [bent at the elbow so that the hand is behind his head]. She was sitting at the window, birdwatching, I suppose, and as soon as I lay down, she came over. She walked across the back of the couch and lay down on the edge of the cushion, up near my head. Then she reached out and set her paw on my arm, just set it there and didn't move it, didn't flex her claws or anything. She just seemed to want to touch me, to make contact. We stayed that way for some time.*

Cat guardians also talked frequently about how their feline companions initiated contact by head or flank rubbing, which is equivalent to the human touch (see also Mertens 1991). This is a greeting behavior among cats, who will often follow up by investigating the person's belongings. A cat guardian describes:

> *When I come home, she's usually sitting there by the chair, whether I come in the car and through the garage or I've been out on foot and come in through the front. She head butts me and winds around my legs, and then, if I've got packages, she's got to inspect them. When I put my things down, she sniffs and sniffs and pokes at whatever I've got with me. She'll then sometimes rub and roll on them. I guess she's saying, "It's mine, all mine!" but whatever it means, it's clear to me that she wants to come and see me and whatever I've got with me.*

Cats will frequently impose themselves on people's activities to make their desires and intentions known. My cat Pusskin regularly paws at my arm for attention when I am working at the computer. Another cat, Leo, watches my husband shave. He also supervises all food preparation. He got so close to the action when we repainted a bathroom that his coat was flecked with paint for weeks. His presence is so reliable that one of his nicknames is "the Helper." Anyone who lives with cats is familiar with how they sit on reading material, making themselves the center of attention. For example, Alger and Alger (2003, 68–70) describe how the shelter cats interfered with volunteers' activities to get attention or make their preferences known.

Evidence of agency among animals helps explain why our experience of them as subjective beings is not solely the result of sentimental anthropomorphism. Our interactions will vary depending on the animal, which suggests that animals shape our response to them in ways that are beyond our projection. If we behaved the same way with every dog or cat, that consistency of behavior would suggest that we see all animals in a similar way. However, because different animals elicit different responses, animals' sense of subjective immediacy becomes apparent. Agency is part of what accounts for this subjective immediacy.

COHERENCE

Whereas agency refers to control over one's actions and intentions, coherence refers to the integrity of the entity to which that agency belongs. Coherence gives agency a place to live, so to speak. If agency provides a sense of self versus other, then coherence provides the sense that the self is a single, bounded, physical entity. As Jerome Bruner and David Kalmar (1998, 311) put it, we acknowledge coherence when "we say of some others that they seem to 'have their act together,' or of our own Self, that some particular line of endeavor is 'very much part of me.'" Several indicators of coherence do not rely on language, making them feasible in non-human animals.

One aspect that identifies a separate, distinct being, human or non-human, is coherence of form. The infancy research confirms that the capacity to recognize others appears as early as two or three months of age. Animals, too, can recognize distinct others. For instance, studies of concept learning show that pigeons can distinguish pictures of kingfishers, which are predators, from pictures of other birds (Roberts and Mazmanian 1988). Although I have not tried showing pictures to my cats, I do know that they regularly distinguish me from others in person. Whereas all the cats come to greet me when I come home, they do not generally approach strangers. Similarly, Shelter volunteers who regularly work with certain animals over periods of weeks or even months find that the cats or dogs will recognize them. One such instance occurred in my work at the veterinary clinic, when I had regular contact with a dog who had had several surgeries and diagnostic tests. I often held him while a technician or vet drew blood, and I sat with him as he woke up after his first surgery. When he recovered and went into the adoption area, other volunteers took him for walks more often than I did. Still, whenever he saw me he wagged his tail, looked in my direction, and pulled to come toward me. One day, when his sight was obscured by an E-collar (to keep him from licking at his sutures), I walked toward him from the opposite end of a long hallway, approaching him from behind as he stood next to another volunteer who was conversing with a third person. As I came nearer, I passed an adjoining hallway and greeted someone else off to that direction. As I did, the dog's tail began to wag and he quickly turned around, as if to confirm that it was indeed me. In other words, he recognized my voice, even when surrounded by other caregivers. Dogs and cats can also recognize the "correct" human form. They will often fear a person in, say, a rain cape or a motorcycle helmet. When Skipper was in the shelter, he once nipped at a kennel worker who suspected that he did so because he (the worker) was wearing large, green rubber gloves while cleaning. The gloves must have been unfamiliar—incoherent with normal human form.

There is ample evidence confirming the capacity for coherence among cats. They recognize specific other cats, which allows them to form friendships and establish hierarchies. One of my cats demonstrated the ability to distinguish particular coat patterns. He had a distinct preference for bonding with "torties," which is short for the mottled black-and-orange coat known formally as "tortoise shell." Similarly, Alger and

Alger (2003, 100) document overlapping friendship patterns among sheltered cats, which indicates the ability to recognize specific others. Coherence has generated the cultural practice of naming animals, which "underscores the animal's particularity—the sense of uniqueness between subjective self and other" (Myers 1998, 71; see also Phillips 1994; Masson and McCarthy 1995). One of the things people do when they adopt is name—or rename—their animals. Although some guardians have a name in mind already, most take time to decide on one, emphasizing the extent to which the name has to suit the animal. As one guardian explained in an interview, "I had to get to know her in order to give her the right name. I had to find out what she was like, and that took a few days." The act of changing an animal's name reflects the degree to which an animal's identity emerges through interaction. For example, a dog whom one guardian had named "Rowdy" became "Sadie" in her new home, indicating vastly different perceptions of the dog's demeanor. Many of the foster cats who pass through my home get new names once they are adopted. When I call the adopters to follow up on the cat's adjustment to the new family, as I usually do, I enjoy learning the new names the cats have been given—a reflection of their new lives.

Research in a dramatically different context confirms the relationship between coherence and naming. In her study of researchers' conduct toward laboratory animals, Mary Phillips (1994, 121) agrees that naming gives individuality to animals. She explains that

> in giving an animal a name and using that name to talk to and about the creature, we interactively construct a narrative about an individual with unique characteristics, situated in a particular historical setting, and we endow that narrative with a coherent meaning.

Proper names, Phillips writes, "are linked to the social emergence of personality, which engenders a matrix of ideas and behaviors unique to one individual" (Phillips 1994, 123). A primary reason that scientists do not name lab animals is that they see them as parts, as opposed to a whole, coherent being. They are sources of cells or tissue, for example, or "containers" for responses and reactions. Thus, even the absence of names reflects a cultural acknowledgment of their relationship to coherence.[5]

Dogs and cats alike indicate capacity for coherence in the act of hiding, which requires a sense of self as a bounded, physical object to conceal from others (see Allen and Bekoff 1997; Sanders 1999). According to Sanders (1999, 137), hiding "shows an awareness that the 'embodied self' is in danger and that concealment is in order." Granted, in hiding during play, the "danger" is non-threatening; the underlying mechanism is nevertheless the same. Cats, having evolved as highly specialized carnivores, relied on the ability to hide in order to hunt. Those mechanisms did not disappear with domestication. As one cat guardian explains:

> *Anyone who has lived around cats has seen this: they hide, they watch, and they attack. The also have very strong notions of when it's O.K. for them to be seen and any cat person knows that cats have got to have hiding places.*

Consistent with this, many guardians reported that their cats would hide when it was time to go to the vet's office. Most often, the cats ran at the sight of the carriers, but sometimes the connection was less clear; the cats somehow "knew" (see also Alger and Alger 2003).[6] More to the point was that in such instances, cats often sought out spots that were not their usual hiding places, which would be the first places for their guardians to look.

Another indicator of coherence is what is referred to as unity of locus. Allow me to use Skipper as an example again. He has a strong reaction to UPS trucks. I say "strong reaction" because I am still not sure whether it is fear or dislike. He recognizes the sound of a UPS truck as well as the sight of one. In an attempt to desensitize him, I took him to a local UPS lot in hopes of rewarding him for being quiet. He recognized the big, brown trucks even when parked. As far as unity of locus is concerned, however, here is the point: When he hears a UPS truck, he looks around for the big, brown truck it should emanate from. If we are out walking and he hears one, say, from around the corner, and a car or a different truck passes by, he does not mistake the other vehicles as the source of the characteristic sound. He looks around. He knows it is coming. My point is that he understands the truck and the sound as a coherent entity that belong together. The sound should come from the truck, and he will not attribute it to a separate locus of origin. Although this may sound basic, it nevertheless contributes to the experience of specifying self and other. Thus, combined with other properties of self, unity of locus helps define self and distinguish self from other.

AFFECTIVITY

A third component of the core self that makes animals' subjectivity available to us is their capacity for emotions. Guardians reported two ways that they could read the emotions of their dogs and cats, and these correspond with two analytical dimensions of feeling.

The first dimension encompasses what are known as "categorical affects." When we think of emotions, what comes to mind most readily are discrete categories of feelings, such as sadness, happiness, fear, anger, disgust, surprise, interest, shame, and combinations of these. As Darwin (1872 [1965]) showed, distinct facial expressions are associated with these basic emotional states. In Darwin's framework, the ability to recognize the conveyed messages of the emotional expressions improves a species's chances of survival.[7] This makes intuitive sense. Recognizing anger in another would lead a person or animal to stay away, and recognizing interest might be what helps young predators learn hunting skills.

All the guardians I interviewed could list emotions expressed by their dogs and cats. Happiness, fear, surprise, and jealousy were among the most common. Sadness and grief were also mentioned. I saw a display of feline grief from one of my cats when another died. The two cats, a male and a female, had slept together, ate and played together, and groomed each other. In short, they had become close friends. When the male had to be euthanized, his companion went through a distinct period of grieving. Indeed, her sadness started before her friend died, when he gradually became withdrawn and uninterested. When the male was gone, the female searched their favorite places for him and stopped eating for a few days. She really did not become "herself" again until we moved into a new house. Granted, the behaviors I characterize as feline grief may not be the same as human grief; this is irrelevant. However, I do know that the female behaved differently after her friend's death than she behaves when she is sunning herself (a state I would call happiness or contentedness) or inspecting my belongings after I return at the end of a day (which I would call curiosity). Along the same lines, Alger and Alger (2003) describe displays of happiness, affection, frustration, irritability, depression, empathy, and jealousy among sheltered cats.

In addition to categorical affects, a second dimension of feelings does not easily fit under these conventional emotional headings. Known as "vitality affects," these are ways of feeling rather than specific, discrete emotions. They are responsible for much of the rhythm that exists in the behavior of humans and non-human animals. Bruner and Kalmar (1998, 311) claim that they "signal the 'feel' of a life—mood, pace, zest, weariness, or whatever." In humans, the perception of vitality affects occurs early in infancy. Even before acquiring language, an infant "comes to recognize and expect a characteristic constellation of things happening" for each separate emotion (Stern 1985, 89). Interest feels one way, for example, fear another.

Long before I knew the term "vitality affects," I knew *about* them. Most of us do, from an early age. When my niece Amanda, now a teenager, was very young, I used to entertain her regularly by making my index and middle fingers into the "legs" of a humorous little character whose body was my hand. I could make this character walk up her arm and tickle her, but more to our mutual delight were the instances in which it did the can-can or the Charleston, say, or a split, or kicked up its heels. This little person—for it did seem to have what it took for personhood—could tremble in fear, hide, take a jaunty walk, or even crawl as if crossing Death Valley without water. Similarly, there is a scene in the movie *Chaplin* in which the main character, bored by a dinner conversation, spears two dinner rolls with two forks. The rolls become the feet of a comparable character who dances and otherwise entertains the guests. These instances work for us because we can read vitality affects.[8]

We can interpret the general state of movement of a faceless hand and know when it "feels" chipper, bedraggled, or lighthearted. This has nothing whatsoever to do with facial expressions, for there are no faces to do the expressing. This is an important way in which the comparison applies to animals. Animals' relatively limited ability to change their facial expressions as dramatically as humans do makes their expressions an unreliable means by which to infer emotional states. As Jeffrey Moussaieff Masson and Susan McCarthy (1995) point out, dolphins seem to wear permanent smiles, but this comes from the contours of their jaws, not their dispositions. Dolphins consequently "smile" even during acts of aggression and when grieving. Likewise, it often appears that dogs smile. However, the vitality affects of animals inform us far more than facial expressions do.

In our interaction with animals, we read vitality affects and thus describe certain individuals as "sweet," "frisky," "serious," "mellow," "hyper," and so on. These are general characteristics of individual animals, part of the core self rather than the expressions of particular emotions. In other words, when we describe an animal (or a person), we usually include some reference to vitality affects. We may say that they are "happy," but in most cases what we mean by that is more along the lines of vitality affects than a discrete emotional state. I found that people do the same with animals. This is an important way that the dog's or cat's core self becomes available to human companions. A guardian who described her dog as "sweet," for example, used that as shorthand for the dog's overall calmness and submissive tendencies. Likewise, a couple who called their cat a "character" used the phrase to evoke his confidence and curiosity, the combination of which often led him into places and situations where more angelic cats would not have dared to tread.

SELF-HISTORY

Self-history, or continuity, is what makes interactions into relationships. Stern (1985, 90) explains that "a sense of a core self would be ephemeral if there were no continuity of experience." The capacity that allows for continuity is memory. Events, things, others, and emotions gain their meaning and are then preserved in memory, in the context of relationships. There are different modalities of memory, and some of these begin to operate very early.[9] The kind of memory required for self-history is pre-verbal, and several aspects of it are apparent in animals.

Self-history operates across the other dimensions of the core self. For example, coherence involves perceptual memory of the form or voice of another person or animal. Affectivity involves encoding memories of what evoked feelings. For example, just as my niece giggled at each appearance of the little character made of my two fingers, dogs and cats get excited when they see a favorite but long-lost toy. In addition, agency requires memory, because the motor plans involved in voluntary muscle movement reside there.

Anyone who has ever taken a dog or cat to the veterinarian knows that animals remember places. For instance, the normally nice dog becomes rigid with tension in the vet's office; Skipper has to wear a muzzle for safe examinations. Leo, the cat who is affectionate at home, hisses and scratches the vet's offending hand. It took two people to hold him so that a third could safely rewrap his injury. Other guardians offered vivid examples:

> *When my dog and I enter the training center, he seems to understand what goes on there. I mean, he is on his best behavior. He stays close by me and engages in eye contact, as if to find out what I want him to do. He knows that it is not playtime. There have been times when trainers have given us designated playtime, and [the dog], who is normally playful, takes a while to relax enough to play with the other dogs in that setting.*

Skeptics will say that something else is going on. For instance, I have heard people say, "Oh, he just smells fear," dismissing the animal's reaction at the vet's as something instinctual. Even if instinct were all that was at work, and I doubt that this is so, the ability to register a particular emotion consistently in a particular setting nevertheless implies a sense of continuity.

One guardian posed an intriguing question: If some animals seem "different" in some settings, such as when "Leo the Lover" becomes a terror in the vet's office, is the animal the same as the one we know at home? In the framework of this discussion, can we really say the animal has a sense of continuity? After thinking about this for a while, I can say yes. The animal behaves the same way in the same contexts: relaxed and playful at home, for instance, but frightened and defensive at the vet's. If it were the case that the animal was only frightened on some visits to the vet, then the case could be made that the animal has little of the kind of continuity I am referring to here. Given that Leo is always frightened and defensive at the vet's, that behavior has become a pattern that the vets and I recognize as "really" him. The obvious objection has to do with what happens when an animal becomes desensitized to something he or she once feared, just as we can learn not to fear hypodermic needles or dental work. Even here, however, continuity becomes apparent. It allows the self to "go on being" but still allows for change as new abilities come on-line and new opportunities arise.

In addition to the veterinarian's office, guardians reported that their dogs remembered favorite places, such as hiking trails. For example, a guardian told me about revisiting a trail after a year's absence:

> *We moved away for a while, and when we got back, I took her to the trail we used to hike all the time. When we'd get there, she'd jump out of the car and sniff, sniff, sniff all over, but she'd run to this one bush. To me, it looked like dozens of other bushes around there, but this was obviously her favorite. Well, the first time we went there after we moved back, she did the same thing: jumped out of the car, sniffed around a bit, and then ran to that same bush.*

The easy way to dismiss this is to say that the bush was appealing to many dogs, and their scents drew this particular dog. Was this the case, or did the dog remember the bush? I asked the guardian what she thought.

> *I think if it were like an entirely new place to her, she would have had to do more exploring, but she remembered that favorite bush.*

I think if she had no memory of the place, she would have sniffed all over. But it wasn't as though she sniffed her way up to the bush. No. She saw it and went to it pretty quickly. She remembered that bush.

Others have documented dogs' abilities to remember places (see Lerman 1996; Sanders 1999). Indeed, Shapiro (1990, 1997) suggests that the lives of dogs are oriented in terms of place rather than time, as ours are. But place is extremely important for cats, too (see Leyhausen 1979; Tabor 1983). The Algers offer numerous examples of how cats demonstrate distinct preferences for particular places to sleep and eat. The same is true for each of my cats. We always know where to find each one during nap and meal times.

Instances in which animals reveal capacities for self-continuity give guardians a sense of the animal as having a concrete history. This history, in turn, gives guardians "the feeling of being the *same* self in relation to the *same* other" (Myers 1998, 73; emphasis in the original). Animals have no sense of today, tomorrow, and next week, but they do remember what happened to them in the past. They have no need for the degree of continuity that gives purpose to human lives. Animals' goals are immediate and embodied. Consequently, their memory skills give them a different capacity for continuity, but the difference is one of degree rather than kind.

PUTTING SUBJECTIVITY TOGETHER

The subjectivity of animals becomes available as relationships offer opportunities to display agency, coherence, affectivity, and history. Memory allows these four capacities to form an organizing subjective perspective. Then, when presented to humans through interaction, the animal's subjectivity confirms the subjective, human other. To illustrate how this occurs, I draw on *Interaction Ritual* (1967, 91), in which Goffman describes the self as "a sacred object." Applying this to human–animal interaction will take a bit of inference, and my doing so might cause Goffman to roll in his grave, but bear with me while I lay the groundwork.

According to Goffman, the "sacred object" of the self is "the product of joint ceremonial labor" consisting of minute, everyday acts of deference and demeanor. Goffman shows these acts to be complementary and overlapping. He gives the example of offering a chair to a guest. The way in which I perform this act of deference will simultaneously say something about my demeanor. I can do it gracefully, clumsily, grudgingly, quietly, or with fanfare. Consequently, this act of deference, as well as the demeanor with which I enact it, will present an opportunity for the guest to display himself or herself as a well-demeaned, deferential person. The guest can ignore my clumsiness, for example, in an expression of good demeanor, and wave off the proffered chair in an act of deference, choosing, perhaps, to sit on the floor like "family." Then I can act in response, and so it goes, as we "hold hands in a chain of ceremony" (Goffman 1967, 90) that takes place unnoticed most of the time. Through regularly performing this "ceremonial labor," we "affirm the sacred quality of others."

Goffman goes on to suggest that interactions and environments allow varying opportunities for the ceremony of the self. Sometimes the ceremony goes well; sometimes poorly. Indeed, we all have known situations in which we could do nothing right, as in "I am not myself today," or the feeling that "This is not 'me.'" Goffman describes the deference-and-demeanor exchange as a "sacred game" and argues that

> if the individual is to play [the game of the self], then the field must be suited to it. . . . Deference and demeanor practices must be institutionalized so that the individual will be able to project a viable, sacred self and stay in the game on a proper ritual basis. (Goffman 1967, 91)

Animals, too, participate in the game. However, I find the concepts of "deference" and "demeanor" unsuitable. I will therefore adapt Goffman's idea to the framework established in this chapter.

The interaction between animals and guardians shows how each senses the subjective self of the other. Both guardian and animal come to the interaction with the necessary capacities for responding to the patterns of agency, coherence, history, and affectivity. Moreover, different people and different animals display distinct patterns across contexts. This allows people and animals, as well, to infer a subjective sense of self in the other. Recall that in Chapter 5 I pointed out that, when someone chooses a companion animal, not just any animal will do. This is because we respond differently to different animals' subjective selves. The differing responses make it unlikely that people can indiscriminately project anthropomorphic qualities onto animals. Although some of this surely occurs, certain animals are simply unable to "carry" certain projected qualities. It just is not "them."

In a revision of Goffman's "sacred game," animals present subjectivity in the various ways outlined in this chapter. Subjectivity gives rise to an unconscious "chain of ceremony" in which the experience of the person and the animal follows the reciprocal lines of, "If he (or she) is this way, then I can be this way." If the relationship is to work, it is because the animal's agency, coherence, self-history, and affectivity create a suitable "playing field" on which the guardian can enact a self—and vice versa. The match is never perfect, just as it cannot be with other humans.[10] In the best cases, the "mismatch" will be optimal—just enough to help to enrich the subjective experience of both the guardian and the animal. In Chapter 5, I suggested that the self thrives when our interactional abilities are challenged. As Myers explains, "This is just what animals offer as interactants: new information—incongruities, interruptions of expectation, challenges—in the context of familiar otherness. . . . As *interactants*, animals present both important continuities and important discontinuities from the human pattern" (Myers 1998, 78–79; emphasis in the original).

Animals are simultaneously like us and not like us. Like us, they can initiate action. They have a sense of what their worlds and the other beings in them should look like. They experience emotions; they are not Cartesian machines. They have histories, and although they cannot narrate them, as we humans can, their inability to do so makes it impossible for them to send mixed messages. As a result, animals make refreshing companions—consistent, yet full of surprises. One guardian summed up the pleasure well when she said, "I really like living closely with creatures whose ways of being in the world are so different from my own."

The point of this chapter has been to show that there is something to animal selfhood. The feeling that animals offer something with which to "connect" is not simply the result of sentimental projection. The animal's capacities for agency, coherence, self-history, and affectivity coalesce, with memory helping to integrate them. Together, these give the animal an organizing, subjective perspective, or a core self. The experiences of the core self concurrently makes recognition of core others possible.

Interaction with animals offers many opportunities to capture evidence of these features. If we think of self as a system of experiences having the features of agency, coherence, affectivity, and history, then our interaction with animals will reflect our perception of these features. For example, when I am working at my computer and Pusskin stands on her hind legs, reaches up, and touches my right arm, I understand her to be wanting attention—sometimes, but not always, in the form of food. This communicates to me that Pusskin has desires and intentions. Moreover, it lets me know that she sees me as one who is capable of fulfilling her desires and intentions. If I ignore her, she will persist until I pay attention to her. She sometimes moves to get a better angle, or she even jumps on the desk to block my view of the monitor. This lets me know that she is aware of her desires, such that she will pursue them until they are satisfied. In demonstrating her capacities for agency in interaction with me, Pusskin confirms my sense of myself as an agentic being.

If our cat Punim enters the room while Pusskin is trying to get my attention, things will continue as before. However, if Leo enters, Pusskin will stop, focus on him, adopt a defensive body posture, and hiss if he comes too close. Leo will try to bully Pusskin on occasion, whereas Punim and Pusskin are, if not exactly friends, certainly amiable co-residents of the same house. In short, Pusskin can distinguish Punim from Leo and adjusts her behavior accordingly. If I were to observe Pusskin without seeing which cat entered the room, her body language would tell me whether Leo or Punim had arrived. Her capacity to recognize coherence of form confirms that sense in me.

When I play a game of fetch with a dog, whether one of my own or one of The Shelter's dogs, I recognize the dog's experience of joy and enthusiasm. That recognition confirms my own emotional experiences. Although I am nowhere near as joyful or enthusiastic about the game as most dogs are, I nevertheless share the pleasure. In short, recognizing the dog's affectivity provides confirmation of that experience in me. Likewise, when I glance up in writing this and see that two of our cats are napping in their usual spots by the window in my study, that continuity confirms my experience of being in relationship with them.

The sense of self as experience fulfills the goals of the self we saw in the interaction in the adoption areas. For instance, recall that the self requires relationships with others. Relationships posit the continuity of others, as well as the expansion of interactional skills, which in turn posits a certain degree of challenge and complexity in the relationships. Because animals have (or are) subjective selves, they can provide the challenges necessary for the continuity of the self. Granted, the complexity of interaction is not what occurs with other people, but it differs in degree rather than in kind.

The idea of self as experience shows how distinct senses of self and other become present in animals. It also shows how an animal's sense of self becomes present to us during interaction. However, this is an incomplete picture of selfhood for animals, as it would be for humans. The animal and the human can also share thoughts, intentions, and feelings. The result is an experience of self with other, as opposed to self as distinct from other. The mutual creation of self with other, or the experience of "we," is the topic of the next chapter.

8

Self with Other: Intersubjectivity

There is a way things look, taste, smell, feel or sound to an animal, a way of which we will have no idea as long as we insist that the only things worth knowing about are our own social constructions of the world.

—BARBARA NOSKE (1997, 160)

The woman sat at her kitchen counter with a fresh cup of coffee in front of her. She picked up the phone and called her HMO to resolve a billing dispute. The task had stretched out for so long that it had begun to feel like a full-time job, and a dead-end one at that. Each time another "customer service representative" put her on hold or transferred her call to yet another department, she grew more frustrated. Each new voice required another telling of the story, another reading off of dates and policy numbers. After listening to the annoying on-hold music for what seemed like hours, she finally spoke to a man who, she thought, could help her. That turned out not to be the case. After repeating the same complaints that she had already made to several other people, she raised her voice at him in anger. When he, too, put her on hold, she reached the limits of her patience and slammed the receiver into its cradle.

The cat, who had observed this from where he sat crouched on the counter about a yard away from the woman, arose and walked toward her. He leaned close to her and sniffed her mouth and nose. Then he reached out and set his paw on her arm. She looked at him and, distracted from her anger, ran her hand down the length of him from head to tail. He raised his hindquarters to meet her hand. After a few long strokes, he sniffed her face again, turned, and, reassured, went back to where he had been crouched.

I asked guardians to tell me about times and situations when animals seemed to understand or share their feelings. The woman who told me this story believed that her cat, accustomed to her usual patience and quietness, was concerned about her anger. Once she had reassured him with her touch, he relaxed again. Most guardians had stories of this sort to tell. For example, a man told me how, when he naps on Sunday afternoons, he stretches out on the couch, and his dog lies down on his nearby bed. If a loud noise awakens them, the dog glances nervously from the man to the window and back at the man again. "If I stay lying down, then he does, too," the man said. "But if I get up, then he's up with me." The dog looks to him, the man said, "to see if everything is all right." Similarly, in Myers's (1998) study of children's interaction with animals in a nursery school, he found that "the children mirrored vitality affects displayed by animals, such as the excitement of the dog, the lethargy of the ferrets, and the calmness of the snake.[1] Similarly, when the guinea pig was restless, some of the children became restless while feeding it" (Myers 1998, 90).

Examples such as these portray animals as having the ability to share intentions, feelings, and other mental states with their human companions. We experience animal others as having the capacity for shared interaction, despite the lack of a shared language. The woman in the opening example knew that her cat needed reassurance, perhaps fearing that she was angry with him. The man whose dog looked to him when startled awake understood that the dog found him a reliable guide to his own emotional state. For people who live closely with animals, this type of knowing is important in the mutual creation of selfhood. By interpreting the content of other minds—human or non-human—we develop a sense of self-in-relation. The core selves of animals, shown through their senses of agency, affectivity, coherence, and history, acquire another dimension when interaction reveals their capacity to share thoughts and feelings with us. Although we humans put our accounts of this capacity into words, the experiences themselves do not depend on language.

The discussion begins by examining three indicators of intersubjective relatedness. These include the sharing of intentions, sharing the focus of attention, and sharing emotional states. In the second part of the chapter, I return to the question raised in Chapter 1 of why people have animals in their lives. The discussion considers how experiences of intersubjectivity with animals shape our identities. In particular, I argue that because animals have selves and we are able to achieve intersubjectivity with them, they help us define situations and enact roles. Animals also enrich our selves through the pleasure that comes from interacting with them, especially from meeting the interactional challenges they present.

SHARING INTENTIONS

Some of the best examples of shared intentions occur during play between humans and companion animals. Because play involves behaviors that can have dramatically different meanings in other contexts, such as mating, fighting, or hunting, communicating intention is crucial in play.[2]

According to my interview data, much of the appeal of companionship with dogs is that they offer people excuses to play. All the guardians I observed or interviewed engaged in regular, vigorous social object play with their dogs, as in games of fetch and tug-of-war.[3] To be sure, this varied with the age of the dog, but even those who were canine senior citizens still enjoyed an occasional frolic. I found that a majority of guardians regularly took their dogs for social play with other dogs at a dog park. Many guardians and their dogs were regulars who showed up every day, or nearly so, early in the morning, at lunchtime, or after work. Weekends, too, had groups of frequent players at the dog parks.

However, dogs and their guardians do not have a monopoly on fun. Among the cat guardians whom I observed or interviewed for this research, about a third played with their adult cats regularly—again, this depended on age, because all of those who had kittens played with them.[4] Some guardians reported that the cats entertained themselves (which might add to the popularity of cats), and in multi-cat households, they engaged in social play with one another. Play is clearly important to dogs and cats throughout much of their life span, and is a frequent form of human–animal interaction.

As I mentioned, play involves behaviors that in other instances have other meanings. For dogs and cats, play involves growling, tackling, mounting, hissing, biting, and scratching. This makes it impractical to define play by behaviors. However, animals can distinguish between, say, fighting and play fighting. Something makes the two contexts different. But what is that "something"?

A definition of play would help, but a definition that pins play to behavior offers none. An alternative is to look at what play accomplishes. The literature on play often uses definitions in which play activities accomplish some function (see Burghardt 1998), such as sharpening skills that might later be involved in hunting or other adult behaviors. However, play does not necessarily have to serve a purpose, either at the time or later in life. As Bekoff and Allen (1998) point out, functional definitions are problematic because of the lag between when play occurs and when the corollary behavior appears in adult activities. This "ontogenetic gap" makes for uncertain correlations between play behaviors and later life consequences. In studies of feline behavior, for example, there is "little, if any, evidence" that play is a form of practice for adult activities (Martin and Bateson 1988, 14). In this light, Bekoff and John Byers offer a non-functional definition. They describe play as "all motor activity performed postnatally that *appears* to be purposeless, in which motor patterns from other contexts may often be used in modified forms and altered temporal sequencing" (Bekoff and Byers 1981, 300–301; emphasis in the original). Purposelessness highlights the extent to which play is a state of its own. Because play involves behaviors that serve other purposes at other times, the important thing is to learn what makes animals play at particular times.

In Goffman's terms, play constitutes a "frame" of experience. "Frame" refers to "a situational definition constructed in accord with organizing principles that govern both the events themselves and the participants" (Goffman 1974, 10–11). Play is a "protective frame" (Apter 1991) or a psychologically "enchanted zone" between players and the serious world. The frame allows the same behaviors to constitute play in one instance but not in another. For example, it accounts for how the "work" I do at my desk can sometimes feel like "play," as well as how Skipper's fierce-sounding growls during our games of tug-of-war do not frighten me.

If play is a particular state, then both players must be able to recognize it. Play thus requires a capacity for intersubjectivity. The play partners must communicate the intention to enter into the protective frame, or signal that they do not wish to do so. With other humans we can say, "Let's play!" However, non-human animals, too, have ways of indicating that they are—or wish to be—playing (see Bekoff 1977, 1995; Bekoff and Byers 1981). As Allen and Bekoff (1997, 99) explain:

> To solve the problems that might be caused by, for example, confusing play for mating or fighting, many species have evolved signals that function to establish and maintain a 'mood' or context for play. In most species in which play has been described, play-soliciting signals appear to foster some sort of cooperation between players so that each responds to the other in a way consistent with play and different from the responses the same actions would elicit in other contexts.

The best example of a play-soliciting signal is the dog's "play bow" (see Bekoff 1977, 1995; Bekoff and Allen 1998)—that is, when a dog suddenly crouches down on his front legs with his back end high in the air. When I asked guardians about play during interviews, or observed them at play with their dogs, most of them indicated—without prompting from me—that they knew their dog was saying, "Let's play!" when he or she did a play bow, although few used this term. Indeed, several people even gave me a demonstration of the play bow to show me exactly what they meant.

Individual dogs add their own embellishments to the basic gesture of the play bow. For example, Skipper ducks his head and spins; Dolly stamps on the ground with a front paw while "chuffing," which is what I call the short growl she uses only in this situation. With the play bow (and its variations), both dogs signal to each other how they intend the following interaction to be taken. Moreover, as Bekoff explains, dogs not only use play bows to initiate play, they also use them in the course of play, communicating, "I want to play despite what I am going to do or just did—I still want to play" (Allen and Bekoff 1997, 103).

Regardless of species, play is a highly complex activity that involves coordinating activities with others, even to the extent of deception. Bekoff's work on dogs establishes their capacity to 1) have intentions; 2) understand that others might misread those intentions; and 3) communicate their intentions to ensure a mutual understanding of the context in which their actions are to be understood.[5] So much for Mead's "insignificant" conversation of gestures. To a dog, a growl does not always have the same meaning.

People who live with cats will recognize when a cat is playful, although to the best of my knowledge cats have no equivalent of the play bow. Nevertheless, the guardians I interviewed could tell when their cats wanted to play—or did not. For example:

> *He sits there looking hyper-alert, and his tail swishes back and forth. Then I go and get the Cat Dancer [a popular cat toy] and we play for a while. He usually quits first. He just walks off, and I know the game's over.*
>
> *When she has that look on her face, she's ready to play.*
>
> *She has a way of walking that says playtime is over. She's bored or just ready to do something else.*
>
> *When we're playing, he'll sometimes stop to lick or scratch himself. I've seen him with mice and bugs, and when he means business he doesn't stop to scratch.*

Similarly, one of the guardians interviewed by Alger and Alger (2003, 21) described how her cats brought her socks and other things that they had previously stolen, which signaled their desire to play.

In social play with one another, cats seem to use trial and error. One cat, feeling playful, will stalk and pounce on another. If the partner also feels playful—or, at least, willing—then the chase is on. If not, one cat will signal clearly that he or she is not in the mood. Hissing, growling, and boxing will tell the playful cat, "Leave me alone!" For example, at the time of this writing, two of the four feline members of my family have different ideas about social play. The fifteen-year-old cat, Pusskin, often rebuffs the play advances of Leo. Leo sneaks up on Pusskin; when she sees him, she flattens her ears back, hisses, and looks off in another direction. Leo shrinks back and lies down, usually on his side, as if to say, "Oops! Sorry. I didn't see that you were busy." This not only suggests that cats can signal intentions to one another, but that they can also read the indicators of their own failure to share intentions.

Among humans, the capacity to share intentions, which appears between seven and nine months of age, is a protolinguistic marker of intersubjective relatedness. The infant—like the animal—already has a core self, shown through agency, affectivity, coherence, and history. The core self allows the infant and the animal to distinguish self from others. The capacity to share intentions adds to this the recognition of other minds. More important, it adds the possibility of "interfaceable" other minds (Stern 1985, 124). This implies that two individuals—whether same species or not—can share the same ideas. It speaks of a different sense of self in addition to that of the core. Whereas the core sense of self appears in overt behaviors—as in learning to sit or not to urinate in the house—the intersubjective dimension points to the possibility that others can share our inner experiences and intentions. It should not be surprising to find this capacity among social animals.

Play is not the only arena in which evidence exists that animals share intentions with their human guardians. Those I observed and interviewed claimed that their cats shared intentions with them around feeding. Several reported that their cats regularly awakened them at a particular time by pressing a paw on their face, arm, or any available skin.

The cats keep this up until the guardians get out of bed and feed them. That this is communicating intention ("Get up and feed me!") rather than simply influencing ("If she gets up, maybe she will feed me") becomes clear in light of several qualifications. First, if the cat were merely influencing the person to get up in the hope of eventually getting some food, the behavior would end once the person got out of bed. Consider this guardian's account of her cat's morning behavior:

> *She will begin to paw at me around 5:30 usually. I can cover myself up and get a few more minutes of sleep, but she really becomes insistent. She starts digging at the covers and makes it impossible to sleep. I've tried putting her out of the room, but she scratches at the door. So I've learned to just get up. Once I'm up, she follows me around to make sure I feed her. She follows me into the bathroom and stays close, usually rubbing against me. Then when we leave the bathroom, she starts off down the hall toward the kitchen, glancing back at me as she goes to make sure I'm coming, too. If I start to make coffee or anything before I feed her—well, forget it. I've learned. She winds around my arms and it's just useless. So I get up, feed her, and then go about my morning.*

Notice that the cat stays with the guardian, continuously communicating her intention until fed. This persistence is evidence that the cat is aware of the goal she intends to obtain. My cat Pusskin persists in a similar way. Our cats eat twice a day, and although another cat initiates the morning routine and the dogs then join in, Pusskin notifies me when it is time for dinner. I mentioned in Chapter 7 that she stands on her hind legs next to my chair when I am at my desk and reaches out with her right paw, the left one steadying her on the seat of the chair. She strokes my forearm. I pet her, futilely hoping, even after all our years together, that she only wants attention and does not really mean for me to stop my work. After a bit of petting, I go back to what I was doing and immediately receive two more strokes on my arm. This time, I pick her up and place her in my lap. She climbs up on the desk and stands between the monitor and me. She reaches out again, now stroking my hand or my shoulder, whichever is closer to her. Because it has now become impossible to work, I get up from my chair and start out of the room. She dashes ahead of me in the direction of the kitchen, looking back at me in the same way that the cat in the earlier example did. Our walk to the kitchen summons the other cats, although I cannot say how they know that, of the many trips I might make to the kitchen in a given afternoon, this one means food.

SHARING THE FOCUS OF ATTENTION

This dimension of intersubjectivity indicates that what you are looking at matters enough to me that I want to share it. In infants, the capacity to look alternately at the mother's face and a "target" that the mother is pointing to or looking at is another pre-verbal sign of intersubjective relatedness. When the child begins to follow the mother's gaze while "checking in" with her eyes and face, doing so suggests more than simply the ability to follow the mother's line of vision. It "is a deliberate attempt to validate whether the joint attention has been achieved, that is, whether the focus of attention is being shared" (Stern 1985, 129). Joint attention requires the capacity for considering alternative perspectives within a shared context of meaning.

Dogs rely heavily on eye contact and the gaze. As Sanders (1999, 144) points out, they "display considerable interest in human facial expression and direct their own gaze in the directions indicated by human attention. Mutual directing of gaze is a display of mutual attentiveness as well as a means by which dogs and other companion animals indicate their understanding of relative status." His field notes describe the importance of the gaze in a manner that will ring true for those who interact regularly with dogs:

> When Emma and Isis look at me they usually pay attention to my eyes. I have noticed on walks how important looking is to them. . . . If on the walk I stop and look in a particular direction, they will stop, glance at me, and gaze off in the direction I am looking. This

> seems a fairly clear indication of their elemental ability to put themselves into my perspective. In a literal sense they attempt to assume my "point of view." If I look at something they conclude that it is probably something important. (Sanders 1999, 144)

Dogs will also initiate the gaze. Several guardians I interviewed said that their dogs told them when it was time for a walk by looking first at the leash or the door, then at them. Once the dog had their attention, he or she would glance back at the leash or door, clearly assuming that the guardian's gaze would follow. Others described how their dogs alternately gazed at them and at a box of biscuits or the cupboard that contained the food.

I noted fewer instances of the mutual directing of gaze among cats and guardians. Indeed, one of the things that some people find intimidating about cats is that they will stare right at you, even when you look away. More likely, they continue to look because you have looked away; their visual skills and their prey drive hone in on motion. Ailurophobes would be better off learning to stare back at cats. I did find, however, that many guardians attempted to share what their cats were looking at. Cats will often gaze fixedly at something so small or so quick that it escapes our attention. Guardians reported that, after noticing that a cat was entranced by some object, they tried to find out what it was, usually without the cat's noticing. A moth gets in the house, say, and provides instant fascination for the cat. The human may not notice the moth at first but cannot fail to notice the cat's impressive vertical leaps against the wall or the sliding glass door—or, more subtly, notices the cat staring up at the wall or the door. The human then ascertains the object of the cat's attention, but the cat remains oblivious to the human. This is not truly "shared" attention, because the cat does not acknowledge the human's interest; we shall soon see an example of this. Nevertheless, the act of attempting to share the cat's attention establishes "an intermediate but still compelling position for animals as social others" (Myers 1998, 95).

Once I began looking for signs of sharing attention among cats, I considered that, given their specialized visual skills, perhaps eye contact and shared gaze were not where we would find the evidence. I began to look for other indicators. I found that cat guardians reported a version of sharing attention that once again concerns food. All those who had had cats for more than one year described something along the lines of the observations I recorded in my field notes during a visit to a guardian's home:

> *[The cat] goes over to the cupboard where the food is and just sits, very upright. We both look at him, and [the guardian] explains that if she doesn't "get it" right away or tries to postpone the inevitable, he will remind her. We talk and wait for a few minutes, trying not to look at the cat and hoping to bring on what she has told me about. In under a minute, he walks over to her and does what she calls his "full body press" [flank rubbing]. Sure enough, after only a minute or two, over he comes. He does "the press" and then walks back over to the cupboard. I suggest that we wait a few more minutes and see what happens next, and I am glad I did because he then does "the press" on the cupboard door and walks back over to [guardian's] leg, where he "presses" again before returning to his position in front of the cupboard. He sits and stares at us, glances at the cupboard, and back at us.*

I will clarify what this indicates. If I point to something, and you look to see what it is, that indicates that you can find my focus of attention. However, if you look at the object to which I am pointing and then look back at me to confirm that we indeed share the same focus, this is another matter. It indicates the realization that we are of two minds, the focus of which can differ but can also be brought into alignment and shared. This may not be a self-aware process; it probably occurs outside of consciousness. Nevertheless, the "checking back" at the other indicates a desire to know and be known.

Cats also vocalize to share attention. In more than twenty years of living with cats, I have listened carefully to their sounds and learned that they use particular vocalizations in particular contexts. They have a sound they make when they see birds; I have always thought it sounded like the bleating of a lamb, but I have heard others refer to it as a chirp. If I look out the window and make the feline "bird call," Pusskin will come to see what I have found. On several occasions, she has looked out the window for the bird and, seeing none, will sniff my face (for some reason) and walk away.

SHARING EMOTIONAL STATES

Most of the guardians I interviewed believed that their cats and dogs were sensitive to their moods and feelings. Other studies confirm this, as well (see Collis and McNicholas 1998 for a review; see also Alger and Alger 2003; Thomas 1993; Masson and McCarthy 1995; and Masson 1997). Although there can be no definitive proof where feelings are concerned, even among humans, there is nevertheless evidence that animals can share some emotional states with us. We humans communicate more than we realize with our faces and our bodies, as well as with chemicals that only animals can sense. Much of the time, these "signals" are so embedded in other behaviors that it is hard to separate a "pure" example. Perhaps the classic illustration is the case of Hans, the clever horse. Clever Hans lived in Berlin in the early twentieth century. He became a celebrity for his purported ability to solve mathematical problems. His owner would ask him for the sum of two numbers, and Hans would give the answers by stroking his hoof on the ground. Many people suspected fraud and accused Hans's owner of giving the horse cues for when to stop stomping his hoof. A commission of highly respected scientists investigated the case and failed to find any evidence that Hans received cues from his owner (see Allen and Bekoff 1997, 25–26). However, a later investigation found that Hans was indeed responding to cues, but of a sort different from what anyone expected (see Pfungst 1911). Hans was picking up subtle, unintentional cues from the people around him, who imperceptibly relaxed or quietly exhaled when he reached the correct answer. Ever since, references to the "Clever Hans Effect" have dismissed unexplained abilities among animals because Hans was *only* reading body language. However, this is missing the point, which is not that Hans was incapable of doing addition. The point is that he could understand the subtle cues that tell horses and other animals—including humans—what they need to know.[6]

Dogs, as pack animals, are highly sensitive to the emotional states of others. Among their wild relatives, survival depends on being able to read the others in the pack. Dogs rely on the same skills to know where they stand in a social group, regardless of whether it is composed of other dogs or human beings. They study eyes and facial expressions, and they receive other signals, such as scent, as well. Elizabeth Marshall Thomas recalls a day that, although she was trying very hard not to show her dark mood, a dog nevertheless sensed what she had managed to hide from people. "Over the great distance he stared at me a moment," she writes, "as if to be sure that he was really seeing what he thought he was seeing, and then, evidently deciding that his first impression had been accurate, he drooped visibly" (Thomas 1993, xvii). Likewise, Bekoff describes his dog Jethro's reaction to his mood:

> Once I came home after a disastrous day at work. Normally, Jethro greets me right away and wants to be fed immediately. But this time he was completely subdued. He seemed to know I needed looking after. He was very compassionate, very empathic, leaning into me, looking at me as though he was thinking, "Marc needs help." (As quoted in Schoen 2001, 172)

The accounts by Thomas and Bekoff call attention to the way dogs not only read our emotional states but also adjust their own according to ours. For example, consider this excerpt from my field notes:

> *Skipper and I walked past a place where workers were tearing up a section of the street to do something with the sewer. They had some noisy machinery there, and there were big chunks of asphalt peeled back and piled up. As we approached, Skipper tensed up and moved closer to me. Then he pulled back and tried to turn around rather than walk past. It was too scary for him. But I kept walking as if everything was normal. Skipper kept glancing up at my face and*

> *then back at the situation. I pushed my sunglasses up on top of my head so that he could see my eyes. I held his eye contact, even while he alternately glanced over at the workers. I kept saying, "Look at me, good dog. That's it. Yes. Good dog, good Skipper," as we walked. He was cautious for the first while, but he quickly locked his gaze with mine and we got past.*

One could argue that this is simply imitation, or "affect contagion." People tend to smile when they see another person smile and to choke up when they see someone else cry. Studies show that infants as young as two months of age will begin crying when they hear tape recordings of crying—even their own (see Stern 1985). Affect contagion exists in the more highly evolved species, and perhaps Skipper was just imitating my calmness in a similar, genetically encoded way, much as he howls when he hears another dog do so.

Imitation may well explain the biological origin of the phenomenon I am describing, or perhaps the mechanism through which it occurs, but it does not go far enough. First, if this were simply imitation, Skipper would not have checked in with me the way he did. By doing so, he indicated that he attributed to me the capacity to have and to signal a feeling state relevant to his own. In this uncertain situation, in which he could have approached or withdrawn, he looked to me to resolve his uncertainty. This is known as "interaffectivity," which refers to "an emotion that is shared and *understood* to be so" (Myers 1998, 90; emphasis in the original). It may well be the "first, most pervasive, and most immediately important form of sharing subjective experiences" (Stern 1985, 132).

Second, the question remains whether we should dismiss Skipper's interaffectivity as "only" instinct or conditioning, or whether it constitutes evidence of a self. In other words, is Skipper just bound to look to me because he sees me as pack leader? Alternatively, does he look to me because he knows, at some level, that I know what he is feeling? Both explanations seem right. Skipper, like all dogs, is genetically predisposed to look to a leader. However, as a social being, his looking to a leader reveals something about the role of relationships. Skipper looked to me not because he had no choice, but because we have a history. If he were walking down the street with another person, he would not have behaved the same way. He and I have shared intentions, objects of attention, and other emotional states. This history reinforces itself and reinforces the self. Intersubjectivity increases feelings of security and attachment to others. Recalling the goals of the self, intersubjectivity packs a tremendous punch in terms of survival value. In short, all evidence points to Skipper's actions as having to do with the self.

Consistent evidence of shared emotional states comes from touching. Dogs and cats alike will initiate contact with humans, and the interaction that follows is mutually enjoyable, as this interview excerpt reveals (see also Sanders 1999, 11):

> *Most nights, we have a routine. After I do the dishes, we watch a little TV. I sit on the couch, and [the cat] jumps up next to me. I have the brush right there [on the coffee table], and I start at her head and work my way down. She loves it, and it's hypnotic for me, too. We both relax. It's almost, well, it's very intimate, too, because we're both trusting each other not to do anything that would hurt.*

Because genetics prepare dogs and cats for different kinds of social interactions, the two species do not have identical intersubjective capacities. Dogs' and cats' resulting social skills make different kinds of relationships possible, and the qualities of one species will not necessarily appeal to everyone. As one guardian suggested, this might explain much of what it means to be a "cat person" or a "dog person." Regardless of which species one prefers, the emerging relationships often involve a rich emotional dimension.

INTERSUBJECTIVITY AND THE SELF

In Chapter 1, I raised the question of why people have relationships with animals. I found the various answers lacking because each of them attributed our relationships to a single explanation, such as a drive to dominate or the inability to relate to other humans. Here, I propose that we continue to have relationships with animals because they become indispensable for our sense of who we are. Identity is not limited to a single answer, or even to twenty answers. It is fluid and, more important, interactive. The identities of sentient animals, both human and nonhuman, depend in large part on relationships with others who possess minds, experience emotions, and have intentions. If humans include animals among these "others," as I argue we must, then the next step is to examine how our relationships with animals shape our identities.

In the following section, I consider two ways that this occurs. First, I discuss how animals become what I call *resources for self-construction.* I argue that animals inform us of our roles and help us define situations, just as other people do. Next, I maintain that interaction with animals enriches our subjective experience. In earlier chapters, I argued that good relationships usually involve increasing levels of complexity. Animals are able to provide this because they are subjective beings and because they show evidence of sharing intersubjective experiences with us. In the final section of this chapter, I look closely at the topic of play to show how animals make possible a level of involvement that enriches our subjective experience.

Animals as Resources of Self-Construction

One of the ways of thinking about the self is as a script or story that connects all the different roles we play in daily life. We might be dedicated workers, devoted parents, and good friends, passionate about some things and indifferent toward others. All of these roles and perspectives are dramatically different, yet something ties them together. Animals participate in the process of self-formation because they call forth different roles, which we then incorporate into our overall sense of who we are.

The conventional way of defining "role" is along the lines of a set of duties and obligations that accompany social statuses. This definition conveys the impression that everyone who occupies a particular status performs in particular ways or risks punishment. To me, this way of understanding roles is too limiting. There are indeed rules about conduct, but our performance of roles is nevertheless highly individual and flexible. Thus, instead of the conventional definition of roles, I prefer to use the term in a sense informed by symbolic interactionism. A role, in this sense, still informs our conduct, but it does so by providing a general perspective instead of a strict blueprint. In short, the symbolic-interactionist approach reminds us that roles do not act, people do, and that the individual tailors the role rather than the other way around.

The symbolic-interactionist perspective on roles includes three closely related ideas: structure, gestalt, and resource.[7] In addition, it assumes that roles occur within situations. It portrays conduct as taking place within contexts filled with other people, instead of suggesting that we act in prefabricated ways independent of the meaning of situations. The interactionist perspective emphasizes that people have an understanding of how the roles in a situation are structured, or what the component parts are in an interactional sense. If I have a doctor's appointment, for example, I can envision the structure of the situation in terms of the roles of doctor, patient, receptionist, nurse, physician's assistant, and so on. Related to this, the idea of gestalt refers to a general grasp of how people behave. In the example of the doctor's appointment, the role of patient makes sense to me within the configuration of the entire situation. I take into consideration the doctor–patient relationship, the initial check-in with the receptionist, the waiting, the interaction with a nurse or physician's assistant, and the handling of payment. The role of patient does not simply demand that I do certain specific things. Rather, the role unfolds within an overall context, or gestalt. The third dimension of role, in which it is resource, refers to the creative dimension of roles. They enable people to do things. Roles allow us carry out various activities jointly with other people. I can interact with a physician because the role of patient gives me the resources to do so. The role of patient enables me to accomplish the goal of seeing a doctor. It does not lock me into particular behavior. Instead, the role allows me considerable leeway within the situation. I use a role as a general basis for acting during the appointment, and I use my knowledge of the general role the doctor will play as a resource for anticipating our interaction.

To apply this to the context of interaction with animals, evidence of their intersubjectivity allows us to include them in the group of others who help us define situations and enact roles. If animals can share at least some of our subjective experiences, then they become separate others who hold definitions of situations against which or along with which we can construct our own. I will illustrate what I mean with two examples. The first occurred before Skipper joined the family, when someone burglarized the house that the cats and I shared. One of the ways that the cats help me to define what is going on is through their roles within the household. For instance, one cat usually jumps to look out the front window at the sound of my car; another is usually the first to greet me, while another, the skittish one, will hide if someone comes to the house with me. Over the years, I defined my typical homecoming in light of the cats' roles. On a particular Saturday afternoon, I pulled into the drive and saw no cat at the window. I opened the door, and no cats came to greet me. This told me that something was wrong; not for a moment did I think that the cats were merely sleeping or had not noticed that I had come home. Because I had shared and trusted the cats' subjective experience, I knew something had happened. Not until I explored further in the house (which I should not have done) did I find evidence of the burglary in progress that explained their behavior. The cats provided the first signs that our shared reality had been disrupted.

The second example is one familiar to all dog guardians: a walk. The situation is defined as, say, a somewhat brisk walk of a mile or so, intended to give the dog a chance to relieve himself and for both of us to get a change of scenery and a little exercise in the fresh air. In keeping with my role of responsible guardian, I put the dog on a leash and have with me a supply of plastic bags to pick up where he leaves off. The structure of the situation includes me, the dog, and potentially other walkers, joggers, and various people and animals we might meet along the way. The role as resource allows me to take this walk in a particular way, to go places where I might not venture otherwise. My gestalt of the situation is not a complicated one, because I know how walks with the dog generally proceed.

However, this information is not enough. Suppose that the dog I am walking is Skipper. Suppose, too, that, as often happens on our walks, we encounter a stranger—a man—coming toward us. Through sustained interaction with Skipper, I know that he is afraid of strange men and will act aggressively toward them if left to his own devices. I have this knowledge in mind when Skipper notices the man and glances up at me. Also through sustained interaction, I know that Skipper is looking at me to find out what to do. I reassure him and get him to do something—walk a few steps in another direction, sit, lie down, or take a treat from me—that will distract him from his fear. In this case, his sharing his emotional state with me helped to create my role.

In contrast, when Dolly and I go for a walk, I have a different role. I again play the role of responsible guardian, which still involves a leash and a plastic bag. However, Dolly has no fear of strangers and no desire to protect me from them. I could stop and talk with the man and allow him to touch Dolly. I could daydream as we walk, whereas with Skipper I must pay attention to things around us. In short, the role I play differs according to the subjective experiences of these two very different dogs.

The devil's advocate could raise the question—and legitimately so—that in both examples I simply know the animals' habits, and this has little if anything to do with intersubjectivity. I would counter this by pointing to the details of our interaction. For instance, although I know Skipper's history of fear of strangers, that knowledge informs but does not determine my role in any given situation. If Skipper had not glanced up at me and given off fear along with his "What do I do?" expression, I would have acted differently. His checking in with me signaled that he understood that I could feel and signal back to him a feeling state that was relevant to his own (see Stern 1985, 132).

Another objection might be that these examples are trivial. Indeed, they are accounts of mundane, everyday experience. That makes my point precisely. I maintain that animals are meaningful for human identity at the level of the everyday and the unremarkable. To be sure, some of us will experience heroic and extraordinary interactions with animals. However, for the most part our relationships with animals will stem from the routine actions of everyday life, such as the greetings, the meals, the walks. It is these things, more than the extraordinary and the heroic, that create the ongoing sense of who we are.

Enriching Subjectivity

Another benefit of living with animals is that doing so improves our own subjective experience. To examine how this occurs, I will focus on play. Animals offer adults rare opportunities to engage in non-competitive play. As adults, we typically play games that have rules and winners. Moreover, we often play to accomplish things that may have nothing to do with the game. Playing golf or tennis, for example, might have more to do with networking with the other players than with the pleasure of the game itself. In contrast, dogs and cats will not admit us to any clubs or put us on any boards of directors. What they can provide, quite simply, is *fun*, which makes for a richer subjective sense of self.

Let me be clear about something from the outset. When we play with our companion animals, we sometimes do so to tire them out. In this case, the activity is *exotelic*, or motivated by an external goal, as when the goal of the golf game is to get to talk to the boss. Here, the play is a means to an end. In contrast, when we are so fully involved and enthusiastic about something that it becomes an end in itself, we are engaged in *autotelic* activity. The term refers to activities we do for their own sake, because the experience of doing them is the primary goal (see Csikszentmihalyi 1990, 117; 1997). An autotelic experience is one that offers the kind of challenge that enriches the self. Examples include playing musical instruments, drawing, painting, or anything that deeply engages us mentally. Although play with animals can be boring—I throw the ball and he fetches it, repeatedly, for example—it does not have to be. It often involves a high degree of concentration; it is an active form of leisure if we make it so—and the trick is to make it so. For example, during the course of one summer, I observed and interviewed a guardian who regularly played with groups of dogs, sometimes approaching ten in number. Here, she describes the experience and her motivation:

> *I don't want to just stand around talking to people. I mean, sure, I'll talk, yeah. But the point of coming here is that it's a space for play. [My dog] and I are here together to do that. Sometimes, when there are lots of other dogs here, it's like we're all playing together. I'm in the game. I'll throw the ball, and a bunch of dogs runs after it, but I don't know which dog will get it, and I don't think they know, either. Then sometimes there'll be a little dodge and chase go on, and another dog will steal the ball. Then they'll all come running back to me at some point, because I have a part in this, too. I have to keep watching the action along with them.*

For this guardian, playing with the dogs required her full attention, and this is the key to enriching subjective experience. A feedback loop develops: If you pay attention to something, you will tend to become interested in it. Once you are interested in it, you will pay closer attention to it. Indeed, you will become so absorbed in it that your attention becomes effortless. In developing a greater capacity for attention, we become more able to control our mental energy. The skill translates to other activities. Our lives become more "ours."

But what is so interesting about playing with animals? At face value, there is nothing to it, as Shapiro pointed out in the earlier discussion. However, Csikszentmihalyi (1997, 128), who has studied autotelic experiences extensively, writes that

> many of the things we find interesting are not so by nature, but because we took the trouble of paying attention to them. . . . As one focuses on any segment of reality, a potentially infinite range of opportunities for action—physical, mental, or emotional—is revealed for our skills to engage with.

With autotelic activities, the important thing is the attitude we take toward it. If we are gardening, for example, to have prize-winning roses rather than for the sheer pleasure of getting dirt under our nails and seeing things grow, the results will not be the same. If we play to tire the dog out, or—as Marc and I did with Punim—to rehabilitate an injury, then some of the pleasure is absent. Once we begin to pay attention to the activity for its own sake, however, we gain the unique benefits of doing so. Play becomes autotelic when it no longer becomes solely something the person does "for" the animal. Instead, it is two partners playing together, as in this interview excerpt:

> *When we first got our cat, you know, my husband hadn't had a cat before, and so he hadn't seen how they get really into playing. We had gotten a package, and it was banded with that plastic stripping that is really hard to tear. They put it around bundles of newspapers and stuff. Anyway, I had a long piece of that. It was bent at one end like an "L" and I was using it to play with the cat. I'd push the end of it out from behind the leg of the table, and [the cat] would come and tackle it. I dragged it all around, up, down, over things. My husband was watching this. The cat and I were totally into it. I have no idea how long we were playing. For me, it was all about tricking him and getting him to follow that darn plastic thing. [My husband] couldn't believe it. He said, "I don't know who's having more fun here—you, the cat, or me watching the two of you." But what he said about me and the cat, it's true. I just get lost in playing with him. It's the purest fun I have. It's completely innocent. There's no competition.*

Another important element of play is that it involves us in interaction with other beings. We benefit from this in several related ways. First, because the play occurs regularly enough to become part of the structure of the relationship, it provides the basis for continuity of the self. It exercises the skills that enable us to maintain and initiate relationships and to "go on being." Second, animals provide "safe" partners with whom to express enjoyment. Like dancing or singing with young children, you cannot do it poorly. For instance, one young woman discussed how, with a dog, there is no such thing as a "bad throw":

> *My dog is so forgiving, you know. She is very athletic and loves to chase this thing [she holds up a soft, flexible Frisbee, a popular dog*

toy made of a rubber ring covered with nylon fabric]. I can't throw worth a damn, but as long as it moves and she can run, we're in business. We come out here almost every night. I love to watch her. She keeps bringin' it back, bringin' it back. She doesn't care if it hits the tree or wobbles or doesn't even go very far. That's not what this is about for us.

This particular woman worked all day in the corporate world, where she had to monitor everything from her appearance to her e-mail. After work, she changed into her "play clothes" (her words) and went to the park with her dog and a Frisbee. Together, they had fun. It nourished a side of the woman that otherwise would have gone hungry. Notice her use of the word "bringin'." Clearly, she relaxed her standards and felt more "herself" than she did at work, which highlights another benefit of play: It lets people out of the "iron cage," to use an appropriate phrase from Weber (1954). He was referring to the way in which our ability to think of ways to do things ever faster and more efficiently becomes a trap. Kenneth Gergen (1991) offers another term that describes the related feeling of never being caught up, never having time to catch one's breath. He calls it "the vertigo of the valued," attributing the sensation to an increasingly technological world. Numerous scholarly studies propose ways to develop fairer economic practices, and hundreds, if not thousands, of popular books offer advice on how to live lives that are more satisfying. Guardians of companion animals have a clear advantage here. Playing with dogs and cats provides an escape—albeit a temporary one—from the iron cage. Many guardians saw play as something mutually beneficial that their animal companions added to their lives. Here, a guardian who works from home describes how her cat lets her out of the iron cage:

I have a crystal paperweight on my desk, and if I hold it in the light just right, it throws rainbows on the floor. [The cat] chases them, and sometimes he "catches" the rainbow and then seems to wonder why it's on top of his paw when he just caught it. When I'm working, it's a nice way for me to remember what's important. We start playing that way, and there's no goal, no competition. Either one of

> *us can stop whenever we feel like it. Cats are totally in the moment, you know, and what could be more that way than chasing a rainbow? It's a wonderful lesson for me not to take work too seriously.*

In a world that makes spiraling demands on adults, and increasingly on children, it is not surprising that animal companionship has increased. Play with dogs and cats is one of the few outlets that people have for non-competitive fun. In this light, animal companionship is "'adaptive' in the evolutionary sense of the word, since it contributes to individual health and survival by ameliorating the stresses and strains of everyday life" (Serpell 1986, 148). Here, in an excerpt from my field notes, I describe such an experience:

> *November 20, 2000. This afternoon, Skipper and I went to agility class. While the sun was high it made for a beautiful day, but as the sun dropped behind the mountains, it quickly got cold. Skipper and I are beginners at agility. I have no plans to become competitive. I registered for the class because I thought it would be fun for us both. Skipper is smart and nimble and appreciates new challenges. I also hoped that the class would help build his confidence and his trust in me. The agility course requires him to do things for no purpose other than because I asked him. He gets rewards just for doing what I ask. He does really well walking the A-frame, and he has mastered the weave poles. He easily jumps through the hoop. However, he doesn't at all like going through the tunnel or the chute. As one who gets claustrophobic pulling on a turtleneck, I empathize. One of the instructors helped us a great deal. He folded up the length of the chute so that it would seem less daunting. He shortened the tunnel, which is accordion-type plastic. But what it really took to get Skipper to go through was me getting down on my knees and encouraging him with pieces of hot dogs and cheese–his favorite treats. I had to show him what to do, or at least show him that it was safe. So, I squirmed through the plastic tunnel, and I low-crawled through the barrel with the chute attached. The melt from two recent snows had turned the ground into thick*

mud. I had dressed for it, but still, I felt a bit like GI Jane. *My hair was in the way. My hat came off every time I stuck my head in the end of either the tunnel or chute to coax Skipper through it. Eventually, he went through both the tunnel and the chute successfully, several times, and I praised him wildly when he did. He really had a good class. Afterwards, as we walked across the field to my car, I thought I had done well, too. Here I was, a week away from my forty-second birthday, playing in the mud with my dog. I do not think I have ever been that muddy before, surely not in my adult life. Skipper was muddy, too. He stayed at my side and seemed to know that we had done something good together. I was cold, tired, and dirty—nothing that a hot shower would not cure. More than that, though, I was happy. Everything around me became profoundly beautiful. As the sun set behind the mountains, it streaked the sky with pink. I saw a red-tailed hawk roosting in a tree. The air smelled faintly of wood smoke. Ordinarily, I am something of a worrier, but at that moment, I did not have a care in the world. The two of us who do not share a language nevertheless communicated, cooperated, and had fun. But for Skipper, I would have stayed inside today. Skipper keeps me outdoors. He gets me out in the fresh air and keeps me moving. He makes me take breaks. He never seems to notice what people would call "bad" weather, and more than a few times when I have dreaded taking him out, doing so has brought me to moments of extraordinary natural beauty. But for Skipper, I would not have seen an owl in flight or heard them calling.*

How poignant that the ability to reason, which we humans put forth as what separates us from animals—or, more accurately, raises us above them—would create an iron cage from which we now try to escape. How fitting it is that animals hold the key. The sense of self that was long considered an exclusively human capacity is enriched by interaction with animals.

Conclusion: Putting Theory into Practice

In this book, I have conceptualized the selves of animals in a way that highlights the similarities with the selves that we experience as humans. In particular, I have shown how elements of a "core self" and a capacity to share thoughts, intentions, and emotions become apparent during our interaction with animals. This idea challenges the conventional, language-centered concept of the self. The symbolic-interactionist tradition, which is where the sociological study of the self is located, maintains that selfhood is an exclusively human accomplishment that relies on the use of symbols, chiefly language. In contrast, I have argued that, although some dimensions of selfhood depend on the acquisition of language, the core self and the potential for subjective experience exist apart from language. To make this argument, I draw on research on the pre-verbal experience of infants and William James's attempts to understand subjectivity. The resulting two-part model depicts the self as a system of goals, which we pursue through relationships, and experiences, which involves the ways in which we respond to and order the worlds around us. Framed in this way, animals, like people, manifest evidence of selfhood. Interaction reveals features of a "core self" among animals as they manifest agency, affectivity, history, and coherence, as well as the capacity for intersubjectivity.

Much of what I have said in this book points in the direction of a postmodern theoretical stance. The changing attitudes toward animals that include seeing ourselves as animals, too, connected to those who share our planet, has emerged in a social context that some describe as postmodern. The postmodern turn within the social sciences has made room for considering ways of knowing that challenge the positivistic biases of modernism. Postmodern thought has opened possibilities—at least, in theory—to hear the voices of those long silenced by virtue of their positions on the margins of power. On the face of things, then, it seems that postmodernism offers the best hope for a consideration of animal selfhood. However, just as theoretical and conceptual space has opened to this possibility, postmodernism has simultaneously declared that the self as a concept has become irrelevant. Any treatment of selfhood must respond to this claim.

The postmodern critique of the self (among humans, at least) holds that our cherished notion that we are autonomous individuals is simply an illusion. Instead, we have "marketed selves" (Dowd 1991) or "saturated selves" (Gergen 1991) constructed by choosing the products we consume. However, even the choices we make are illusory, because the products are essentially the same and it makes little difference what we choose. Underneath the brand names that we wear and use, there is nothing transcendent, and any such notion that there might be is simply a nostalgic vestige of the Enlightenment. The sooner we realize this and release our death grip on the self, the sooner we can abandon the corollary notion that our lives have meaning and purpose—another Enlightenment delusion. The idea that people could improve the world, even their corner of it, is merely a romantic fantasy, and one that greatly exaggerates human agency.

The postmodern demise-of-the-self argument is theoretically plausible but empirically problematic. Although it is easy to point to the ads, the theme parks, the franchises, and the brand names and say that these—and nothing more—make up the contemporary self, I do not agree. Most of those who say that the self has disappeared have neglected to ask people what *their* experience is. Were they to do so, they would find that the self as experience remains relevant. Examples of research from two different contexts, in particular, support this. The first is Patricia and Peter Adler's study of transience as a lifestyle (Adler and Adler 1999). Briefly, the Adlers studied resort workers who frequently relocate to far-flung destinations and appear to have no "roots" in any traditional sense of the word. Because they must continually adjust to new locations, make new friends, convert new currency, and remain open to new experiences, these "drifter-workers" are prime candidates for a postmodern identity. The Adlers found that, although these postmodern people manifest some aspects of correspondingly fluid identities, overall, their transience "has not resulted . . . in the loss of a core self" (Adler and Adler 1999, 53). As the Adlers explain:

> Theories of the postmodern self focus on changes that have occurred to the more surface aspects of the self, which are recognizable here. But these theories have not looked at the deeper aspects of the self. . . . While many theoretical depictions of the technological and cultural environment arising in the postindustrial world appear accurate, it seems that the postmodernists' most pessimistic view of the demise of the self has not been borne out; rather the core self has adapted to contemporary conditions and thrived. (Adler and Adler 1999, 53–54)

Other research that empirically corroborates the endurance of the sense of self is my own work on members of Codependents Anonymous, people in a unique position to remake themselves according to any available image (Irvine 1997, 1999, 2000). The members had all experienced the end of an important, committed relationship and had the opportunity and the motivation to do things differently. Most were highly mobile geographically, which gave them great freedom to become someone different. I studied them as they told their stories to rooms full of strangers, with no one who could say, "But that's not *really* you," and I interviewed them privately for comparison. I found no evidence of the image-conscious, postmodern social chameleons for whom, as Gergen (1991, 7) puts it, "the very concept of personal essences is thrown into doubt." Instead, I found people who strove for overall purpose and direction, even in the wake of disruption. Their main commitment was to discovering who they "really" were and to be true to what they discovered.

In short, the idea of a "real" self seems inviolable. Although a few postmodern academics may theorize its demise, most people seem to believe it is alive and well. If their experience is merely a delusion, then I must wonder how the demise-of-the-self critics managed to escape to a place from which to observe everyone else being deluded.

The empirical evidence calls for new ways of conceptualizing the self that make it more consistent with experience. This option leads to considering alternative forms of subjectivity (see Holstein and Gubrium 2000). While I would extend this to include those forms experienced by non-human animals, others have already used it to study selfhood in non-Western cultures, long dismissed from the discussion for not complying with the hyperindividualistic standards that allegedly count as indicative of the presence of the self. Clifford Geertz (1984), for instance, examined subjectivity among the Javanese, Moroccan, and Balinese and, although not engaged in the same type of analysis that I have done here, his concern was similar. He encouraged his readers to consider the various manifestations of subjectivity across cultures, and not to presuppose its absence in cultures that do not subscribe to a Western model.

Selfhood becomes apparent among animals in the same manner: as an alternative form of subjectivity. If we begin by assuming that language is necessary, then animals are out of the discussion, as are people with disabilities, injuries, and conditions that leave them unable to speak. By following the language requirement to its conclusion, as I have already mentioned, people who cannot speak have no selves. However, people who work with and care for others who cannot speak will readily acknowledge the selfhood within. Friends and caregivers will "speak for" the mute, the autistic, the brain-injured, the Alzheimer's patient, and the severely retarded. The same takes place, albeit with a few differences, between people and animals. If we look for selfhood in interaction, we will see it even without language. In this way, I have attempted to make animal selves and their influence on the selves of the people around them empirically accessible. I have tried to delineate a plausible notion of animal selfhood—one that builds on capacities present in humans, as well, but also acknowledges that human selfhood is different in degree rather than kind from that of animals.

The idea that animals have selves has numerous implications, and I will discuss some of them here. The first concerns my discipline of sociology. The recognition of animal selfhood will require sociologists to rethink the social world. This means not simply creating the "sociology of animals," but recognizing that the social world is not exclusively a human one. For instance, more than half of American households include cats and dogs, and about 90 percent of those asked consider the animals family members.[1] Yet sociological definitions of the family exclude animals. Sociologists have studied the impact of divorce, the division of household labor, the influence of religion, and several other dimensions of family life, but they have disregarded the animals who are also part of the dynamics. To be sure, sociologists have begun to expand the definition of what constitutes a family, but they have done so while ignoring the dogs and cats whom those they study would count as part of the mix. The study of race and ethnicity can also shed light on the role of animals, because fewer minority households include animals. Other areas of sociology need to bring animals in, as well. Criminologists must consider animal cruelty and abuse as a crime in itself rather than as an indicator of future crimes to humans. In addition, sociologists who study work and occupations have many human–animal teams to investigate, as well as the instances in which people bring their companion dogs to work with them.

Adding animals to the sociological mix will expose many assumptions and threaten many established perspectives. It will enrich our understanding of interaction by extending it beyond the modernist reliance on language. Bringing animals in would also enlarge the discussion of what it means to be social, for animals, too, interact within social contexts. They adjust their behavior to the reactions of others. They do this even with other species, as when a cat knows the intentions of a dog. They can see themselves vis-à-vis other creatures around them, as in the pack society of dogs and the social groups of cats. They have hierarchies and compete for resources. They experience emotions and share them with others. Because such arrangements do not rely on spoken language, sociologists—and especially social psychologists—must find ways to understand and theorize them. As Gail Melson (2001, 190) puts it, "'Ideas about self and others' would expand to 'ideas about self and other beings.' Theory of mind would be reframed from 'understanding people as mental beings' to 'understanding the mental life of other beings.'"

Other implications of animal selfhood go far beyond the scholarly realm. Acknowledging the value of animals' lives in this way will profoundly influence our treatment of them. Suppose, for example, that you agree that animals have the components of a core self and subjectivity that I have outlined here. You agree that animals are self-aware. They are not Cartesian machines, and therefore they have an interest in not suffering. You acknowledge that we owe it to them, as self-aware creatures, not to cause them harm. However, let us suppose that beyond this you are not willing to go. You want to draw a line between human experience and animal experience. You agree that animals have an interest in not suffering, given their self-awareness, but in your view animals are qualitatively different from humans because, for example, they cannot plan their lives and write their autobiographies. They lack the capacity to care whether they live or die. If this sounds plausible to you thus far, you will probably admit that, as long as they remain alive, animals have an interest in avoiding suffering. However, you doubt whether during that life it matters to them whether they are someone's property.

This position is concerned with the quality of animals' lives. It is the "animal-welfarist" position. It is characterized by the belief that, "while humans should not abuse or exploit animals, as long as we make the animals' lives comfortable, physically and psychologically, then we are taking care of them and respecting their welfare" (Bekoff 2000, 43). The welfarist view has its roots in Bentham's famous statement, quoted in Chapter 2, highlighting animals' capacity to suffer regardless of their inability to reason or speak. Bentham's view, radical for its time, held that we had a moral obligation not to cause unnecessary suffering. As a utilitarian, Bentham held that the morally correct action in any instance was that which maximized pleasure for those involved. He argued that, because suffering is undesirable, the moral choice is not to inflict it on any creatures that have the capacity to suffer. Bentham's view gave rise to the "humane treatment principle" that is the basis for animal-welfare laws. The contemporary version of his argument appears in the work of the philosopher Peter Singer. In *Animal Liberation* (1990), Singer agrees that animals have an interest in not suffering. However, they lack the kind of self-awareness possessed by humans; this quality gives human interests preference over animal interests. Thus, we may use (and kill) animals for our purposes, including keeping them as property, as long as we treat them humanely. When we kill them, we must do so quickly and minimize the suffering involved.

If you agree with the philosophical basis of the welfarist view, it should compel a particular course of thought and action.[2] Animals can suffer in many ways, and the obligation not to cause any suffering must go beyond giving a dog or a cat a home. For example, the welfarist position posits abandoning inhumane training practices. It also implies reconsidering the question of purebred animals when breeding perpetuates disease through "aesthetically-based dysfunctional 'breed standards'" (Rollin and Rollin 2001). Anyone truly concerned with companion animals' welfare must advocate the prohibition of practices such as tail docking, ear cropping, declawing, and de-barking.

A great many things can affect an animal's welfare. Surely, direct physical pain affects it, but suffering can also come from fear and distress. Every animal who enters a shelter experiences these forms of mental and emotional suffering. Regardless of how clean and attractive the environment is, the animals are visibly anxious, especially as newcomers. The welfarist response should be to reduce the numbers of animals moving through shelters. The welfarist should thus advocate for prepubertal sterilization. This is especially urgent, given that in some shelters as many as 80 percent of the animals do not "move through" but are killed. Changing this will involve educating people about what it means to take responsibility for an animal's life. As a society, we have come to understand the link between cruelty to animals and cruelty to humans. High rates of animal abuse accompany domestic violence, and childhood abuse of animals often appears as a marker of violence later in life (see Flynn 1999, 2000a, 2000b for reviews). Rollin and Rollin (2001, 9) extend this connection and ask, "If failing to check cruelty to animals inexorably leads to cruelty to humans, does something similar result from failing to honor our responsibilities to animals?" In doing this research, I learned that the answer is "yes." I saw children accompany their parents to surrender a dog or cat because the family was moving or because they could not "deal with" some aspect of normal animal behavior, and I knew what they were learning. As long as the act of "getting rid of" an animal goes without consequence, and as long as animals are cheap, the answer to the Rollinses' questions will be "yes."

Given the high incidence of cat abuse, welfarists should recognize and act on the urgent need for education about feline behavior and humane treatment. The effort to educate people about what guardianship entails will have to begin with children. They must learn from sources other than the *101 Dalmatians*. Their education must begin early, as evidence suggests that children around age seven or eight have the greatest interest in and ability to apprehend the content of such teaching. For example, a team of researchers created a ten-week, home-based wild-bird–feeding program aimed at increasing elementary-school–age children's knowledge about birds (Beck et al. 2001). They supplied families with feeders, seed, and educational materials. At the end of the study, three-quarters of the children showed improved knowledge, measured by ability to identify species and sexes of birds and distinguish those that would be drawn to feeders in the area—for example, the children could determine that a cardinal, not a flamingo, would come to feed. The greatest improvement occurred among children between ages seven and nine. Likewise, a program designed to prevent dog bites, a major childhood public-health problem, showed that the optimal age for educating children about dog-bite safety is eight, and the optimal grades are third and fourth (Spiegel 2000). Clearly, any programs that educate children about their responsibilities to companion animals should be informed by such studies and begin at similar stages (see also Myers 1998; Melson 2001). Children already receive education about avoiding all sorts of harm that *may* come their way. The welfarist perspective would push for education about properly undertaking a responsibility that has a *high probability* of coming their way. In short, the welfarist view must influence how the human members of our moral community treat animals

Suppose, however, that you think animals' experience of self extends beyond a basic awareness of pain and pleasure. Perhaps you think that animals have the ability to feel pain and pleasure for some reason, instead of experiencing those sensations as ends in themselves. The logical purpose for the capacity to feel pain and pleasure is to pursue one and avoid the other, and to go on doing so. It would seem, then, that the creature who can feel pain and pleasure has an interest in life, because remaining alive opens the possibility of further feelings of pain and pleasure.

If you find this convincing, then your position is the animal-rights view, the principal tenet of which is that animals have the basic right not to be treated as things, particularly as the property of others. There are different approaches to animal rights, as well as a great deal of confusion about the potential implications of extending rights to animals. Although I can offer only a cursory treatment here, I will attempt to make it a concise one by focusing on the work of two leading animal-rights scholars, Tom Regan and Gary Francione.[3]

The philosopher Tom Regan, author of *The Case for Animal Rights* (1983), rejects the utilitarianism of Bentham and Singer because it attempts to maximize some "good" without specifying how such "good" will be distributed. Specifically, the utilitarian view admits that some interests are at least potentially more relevant than are others. For example, human interests can outweigh animals' interest because they are held by humans. At the outset, then, utilitarianism does not include the principle of equal inherent value. This principle holds that the individuals under consideration—and for the moment, I will talk about humans—have an unconditional and equal value, independent of their value as a resource to other human beings. Equal inherent value is what prevents us from treating certain other human beings as things. It is "pre-legal" in that it is a prerequisite to other, additional rights, such as freedom of speech and the right to vote. Regan extends equal inherent value to animals because of their status as "subjects-of-a-life," a category that encompasses all normal mammals over a year old. As Regan (1983, 329) puts it, "Like us, animals have certain basic moral rights, including in particular the fundamental right to be treated with the respect that, as possessors of inherent value, they are due as a matter of strict justice." In his view, this means giving equal consideration to animals' relevantly similar interests. It means we cannot devalue animals' interests simply because animals hold them. It means that an animal's interest in not suffering is no less significant than my interest in not doing so. Equal consideration constitutes the foundation of all moral theories in that it guides us to treat like cases alike. We acknowledge and apply equal consideration when we say, for example, that neither race nor religion can justify devaluing the interests of certain people. In advocating equal consideration for subjects-of-a-life, Regan calls for sweeping changes, including the abolition of commercial animal agriculture, trapping, hunting, and the use of animals in research.

The attorney Gary Francione (1995, 1996, 2000) offers a significantly different argument. In his *Introduction to Animal Rights* (2000), Francione agrees with Regan about the importance of equal inherent value. However, Francione argues that attributions of equal value are meaningless because animals are considered property. As he puts it, "Animals lose because their status as property is *always* a good reason not to respect their interests in not suffering. *The interests of property will almost never be judged as similar to the interests of property owners*" (Francione 2000, 86; emphasis in the original). Francione emphasizes the basic right not to be treated as a thing or a resource. This is a precursor to additional rights, and although a complex debate revolves around about which additional rights we might have, most advanced societies uphold this basic right.[4] Without it, all other rights mean nothing. It is, as Francione puts it, "the minimal condition for membership in the moral community" (Francione 2000, 95). The idea of basic rights assumes the corollary idea of equal inherent value, which Francione would extend to all sentient beings. He writes:

> Sentient beings use sensations of pain and suffering to escape situations that threaten their lives and sensations of pleasure to pursue situations that enhance their lives. . . . Sentience is what evolution has produced in order to ensure the survival of certain complex organisms. To deny that a being who has evolved to develop a consciousness of pain and pleasure has no interest in remaining alive is to say that conscious beings have no interest in remaining conscious, a most peculiar position to take. (Francione 2000, 138)

It makes moral sense to extend the right not to be treated as things to animals based on their sentience. Indeed, Western societies have already endorsed this with the laws that reflect the humane-treatment principle, which holds that animals are not things that have no interests. The basic elements of the humane-treatment principle hold that animals at least have an interest in not suffering. At minimum, then, most people already accept that we have the moral obligation not to cause unnecessary harm. To justify this obligation, however, we must apply the principle of equal consideration to animals, for it is illogical to have moral obligations to things. Equal consideration means that animals' interests cannot be judged as less important simply because animals hold them. This, in turn, implies that animals have equal inherent value and cannot be treated as property. This status will entail profound change. As Francione (2000, 127) explains, "Admission to the moral community may not have the same meaning for animals as it does for humans, except insofar as such membership rules out the treatment of any member exclusively as the resource of others." Granting this basic right to animals will mean abolishing, not regulating (as in the welfarist position), institutionalized forms of animal exploitation, including the use of animals as food, clothing, and research subjects. It will also mean the end of pets and even companion animals. For if animals have the right not to be treated as things, then we cannot justify breeding them simply to serve as our companions. In other words, recognizing animal selfhood and its influence on human identity should lead us to acknowledge the value of animals' lives. In turn, we should realize that it is immoral to keep them for our pleasure, regardless of whether we call them companions or pets.

I realize the impact of this claim. As I write, I am surrounded by animals. Two of our four cats lounge in window hammocks over my desk, and both dogs laze close by, waiting for me to do something that involves food or a walk. I cannot imagine my home without animals in it. Yet if they had the basic right I just described, they would not be here. Although I have argued that animals profoundly influence human identity, our wanting to have them around does not justify continuing to breed puppies and kittens. Perhaps the greatest influence they can have on our identity will be to prompt us to act on what we know to be true.

I began this book by examining how culture has influenced our views of animals, and I turn to the same direction in ending it. Today, we hold contradictory views about animals, and those we even profess to love have a dual status as companions and property. Dogs and cats are so prevalent in most people's lives that they practically constitute a birthright, and suggestions that they deserve better treatment often fall on deaf, even defensive, ears. For example, on a recent, extremely hot summer day, I watched a man in his early twenties "tow" a puppy with a bicycle for several blocks—meaning the dog was tied to the bike as the man rode along. When they reached their destination, the man tied the panting pup to the pedestal of a public telephone in the sun and walked off to shop at a nearby store. I approached and asked the man whether he had water for the dog. "Not on me," he said. I told him that I would give the dog some water, and I advised him not to leave his dog there, not only because the dog had to sit in the sun, but also because tethering was illegal and the man could get a ticket.[5] The young man swore at me and added, "I don't need you to tell me how to take care of my dog."

If I (or someone else) had approached the young man differently, perhaps by asking to pet the dog, and had then worked into questions about the man's relationship with the dog, he probably would have said the dog was his best friend. He might even have admitted to loving the dog. Yet he forced his beloved best friend to keep pace with a bicycle on a hot day, without water to quench his thirst. Moreover, he left his best friend on the street, treating him as he would never treat his cell phone or his sunglasses. My "butting in" had the same effect as if I had told the man he ought to wash his car: I had no right to tell him what to do with his property.

The notion of animal selfhood gives us profound obligations, as individuals and as a society. I do not profess to have the answers to all the questions it raises. However, I am certain of one thing that must underlie the search, and that is the intrinsic value of animal lives. Their influence on human identity has been inestimable. It is time to reciprocate by undertaking the difficult and often uncomfortable task of wrestling with the moral dimensions of our relationships with them.

Appendix: Methods

This book draws on several types of data from different sources. For the sake of clarity, I will go through them systematically.

ETHNOGRAPHIC RESEARCH AT THE SHELTER

I spent approximately 360 hours observing the interaction between people and animals at the private, nonprofit humane society that I refer to as "The Shelter." It is a full-service organization, meaning that, in addition to adoptions, it offers veterinary services, education, training, and behavioral consultation, and serves as the headquarters for the city's animal-control services and animal-welfare investigations. It is among a growing number of shelters in the country that do not euthanize healthy animals. However, it is not a "no-kill" facility. Animals are put to death if they have serious, untreatable health problems; if a guardian requests (and pays for) euthanasia of an elderly or ill animal; or if the animals are behaviorally unable to be placed in a home because of, say, aggressive tendencies. A well-run foster-care program provides temporary housing for animals who, because of special needs, would be euthanized at other facilities.

At the time of this writing, The Shelter had a full-time staff of 40 and more than 500 volunteers. In 1998, I began volunteering in the veterinary clinic, and I continue to do so as of this writing, assisting the technicians with pre- and postoperative care. My data collection began the next spring, when I began to work as a dog-kennel assistant a few mornings a week. My duties included walking and grooming dogs, cleaning kennels, working on basic canine manners and obedience, and doing anything else that might help relieve the boredom of shelter life and make the dogs more adoptable. I often worked closely with particular dogs who, because of behavioral problems, had been at The Shelter for several weeks or even months. In this case, a staff member would work out an exercise, training, and play plan for the dog, and other volunteers and I would help execute the plan and write progress reports. Meanwhile, I recorded observations about the practice of sheltering and the interaction I had with staff and animals (see Irvine 2002).

My work in the kennels gradually led me into another volunteer position. Because I was already answering clients' questions about specific animals and providing general information about adoption, behavior, training, and care, I took the next step and became a trained adoption counselor. This involved introducing people to animals they were considering for adoption. During an adoption-counseling session, the counselor tries to determine whether the animal and the person will be a good "fit." In the course of this work, I became curious about people's interaction with dogs and cats in the adoption area. I began taking thorough notes on how long they looked at particular animals, whether they adopted that day or just visited, and what, if anything, they said to the animals or to the other people with them.

Yet another volunteer role yielded more than 150 hours of observation on what I call the Mobile Adoption Unit, a thirty-foot-long recreational vehicle that serves as a traveling branch of The Shelter. Five days a week, a volunteer (such as myself) and a staff member who administers and drives the Unit, take a selection of adoptable cats, rabbits, small mammals, rodents, and a dog to various sites throughout the county. Shopping centers—especially those with supermarkets or national chain stores—libraries, local festivals, and fund-raising dog washes are some of the locations at which the Unit regularly appears. On board, people can adopt animals, donate food or money, and obtain a wealth of information about the care and behavior of companion animals.

The Mobile Adoption Unit spends four hours at a given site. During this time, visitors usually number about one hundred. For many of these visitors, the Unit is their only exposure to The Shelter. Therefore, the work entails a high degree of public relations. My role on the Unit involved—in addition to caring for the animals—welcoming and talking with clients. This meant discussing a particular animal or animals in general, asking for donations, handling adoptions, and answering questions about The Shelter's services or animal behavior. For this research, I took notes about the interaction on the Mobile Adoption Unit in a small notebook.

AUTO-ETHNOGRAPHY

In late spring 1999, I began making extensive notes about my interactions with my own companion cats. Although I have shared my home with multiple cats for my entire adult life, only then did I begin to record the details of our lives together. I made thorough notes about the taken-for-granted activities I engaged in with the cats, such as feeding, play, touch, and eye contact. I also noted how the cats interacted with one another. Then, that summer, I adopted Skipper, who makes regular appearances in this book. Having been a "cat person" all my adult life, I never imagined that I would live with a dog. I assumed the cats would never stand for it and I wondered what I would do with a dog while I was working. However, it seemed that almost everyone in town had a dog. Presumably, at least some of those people had jobs, and some, I imagined, had cats, too. If they could do it, I thought, I probably could. I brought Skipper home, and he brought many new joys—and some hardships—into my life. As the cats and I learned and adjusted to a dog's ways, I took it all down in my notebooks.

INTERVIEWS

I examined my auto-ethnographic notes and those from The Shelter regularly, searching the data for emergent themes and patterns. As the selves of animals emerged as a clear finding, I developed a list of questions that focused on people's attraction to particular animals and the ways in which animal selfhood becomes apparent. With this in mind, I conducted semi-structured interviews with forty people who had adopted and surrendered animals. The Shelter helped me recruit interviewees by attaching an information and consent form to their adoption and surrender paperwork for a month during 2001. I disqualified people who had surrendered litters of "barn cats" to whom they had formed no bonds or had simply surrendered a stray they had found on their way to work. I also disqualified people who surrendered ill or aged animals for euthanasia, because this was outside the scope of my research. In the interviews, I asked how people made decisions to adopt or surrender; how they chose a particular animal for adoption; and how they adjusted to gaining or losing an animal. I also asked about the "texture" of everyday life with the animal in the routines such as play, feeding, touch, and other types of behavior and interaction.

ADDITIONAL PARTICIPANT OBSERVATION AND INTERVIEWS: THE PLAY DATA

As it became clear that play was a frequent form of interaction between people and animals, I sought ways to learn more about it. Thus, another source of data was participant observation at two urban dog parks. These are plots of land open to the public but reserved for off-leash dog play. At peak times, especially after 4:30 P.M., as many as twenty dogs and their guardians use the spaces. I visited one park regularly for four months and the second for six months. I brought one or both of my companion dogs, observed guardians and dogs at play, and took notes as soon afterward as I was able. I supplemented my observational data with field conversations. I talked at length with guardians about play routines, objects, amount of time spent playing, frequency of visits to the park, and their interpretations of their dogs' play behavior. To round out the cat side of the play data, I interviewed cats' guardians (recruited through The Shelter) about their play routines. In several instances, I went to their homes to do the interviews and consequently observed play firsthand (see Irvine 2001).

This research is biased toward people who enjoy animal companionship. I did not intend to do an exhaustive study of the range of human–animal relationships. Clearly, animals are involved in the social world in many ways, all of which need investigating. The more we seek to understand those who share our world but cannot tell their own stories, the better off we will all be.

Notes

INTRODUCTION

1. Saint-Exupéry 1971 (1943), 59.
2. This is consistent with the findings in the research literature, which reveals that the best predictor of having animals as adults is having had them as children. The kind of relationships a person had with animals as a child will also influence the kind of relationships he or she has with animals in adulthood: See Kidd and Kidd 1980; Poresky et al. 1988.
3. I struggled over whether to include this recollection for fear that it might appear to justify petting zoos and having "exotic" or protected species in them (for example, "Petting zoos teach kids to connect with animals; therefore, petting zoos are good."). Petting zoos are usually terrible places for the animals. The elephant I met was most likely an orphan who was sold after his mother was killed. There are far better ways for children to experience the wonder and dignity of animals, and I encourage parents to look for them.
4. Since then, I have put the importance of narrative in a different perspective. I agree with Ulric Neisser, who wrote, "Self-narratives are *a* basis but not *the* basis of identity" (Neisser 1994, 1; emphasis in the original).

CHAPTER 1

1. The statistics are derived from a national survey of 80,000 randomly selected households. A total of 54,240 questionnaires were used in the tabulations, for a response rate of 67.8 percent. The result is a survey of households rather than a census of the companion-animal population.
2. For thorough discussions of the various theories of where dogs came from, see Coppinger and Coppinger 2001. See also Budiansky 1992; Clutton-Brock 1995; and Coppinger and Schneider 1995.
3. Although dogs differ visibly on many dimensions, and they differ from their wild ancestors, tests on mitochondrial DNA reveal an intriguing finding. (Mitochondrial DNA passes from mothers to daughters without recombination, providing a reliable way to research lineage.) Dogs, wolves, and coyotes are far more similar genetically than are the various ethnic groups of human beings, which are classified as the same species (see Coppinger and Schneider 1995).
4. In the late twentieth century, wolves have been reintroduced to many areas of North America, and the protection of wolf populations remains highly contested. For a thorough treatment of the wolf's place in relations between humans and nature, see Emel 1998.
5. James Serpell (1986, 127) raises an intriguing point about the role of technology in pet keeping: "It is surely significant that nocturnal rodents such as mice, rats and hamsters have only become popular as pets since the advent of the electric light; an invention which has artificially extended the period of human activity into the hours of darkness when these animals are normally active."
6. The journalist Stephen Budiansky (2002, 76) writes that "if a trained expert is shown a wolf skull and a dog skull, he will have little difficulty telling the two apart. But when zoologists, natural-history-museum curators, hunters, veterinarians, game wardens, and professional naturalists were asked to distinguish between specimens from a wildcat and a domestic cat, they were right only 61 percent of the time. That's hardly better than what ukulele players, avocado growers, and automobile designers could have done by flipping a coin."
7. This raises an important point about the importance of keeping companion cats indoors in North America. To native wildlife, the cat is an exotic predator. North American birds and other animals did not evolve in conjunction with cats, so they did not develop defenses against them. Therefore, cats who go outdoors have a tremendous advantage in hunting. They have been responsible for declines in numerous bird populations, especially hummingbirds, who are particularly vulnerable because they must hover to feed.

Well-intentioned humans who put bells on their cats thinking that the sound will warn birds are mistaken. Birds have no reason to associate the sound of a bell with impending attack. People who think that well-fed cats will not hunt are also mistaken. Hunger and the drive to hunt originate in different areas of the brain. Thus, a cat who is not hungry will continue to hunt. At the risk of editorializing, a guardian who lets a cat outdoors is setting loose a well-fed, exotic predator. In an article in the prestigious journal *Nature*, Kevin Crooks and Michael Soulé (1999) studied domestic cats allowed outdoors in a residential area in San Diego. Seventy-seven percent of the households let their cats outdoors, and 84 percent of these outdoor cats brought home their "kills." Each cat killed fifteen birds annually, along with nearly double that number of small rodents and lizards. Crooks and Soulé claim that, at these rates, native bird species will soon become extinct. They estimate that 75 percent have already disappeared. The solution is to keep cats stimulated and content indoors by providing window perches, scratching posts, toys, and, most important, companionship. For more information, see the website "Cats Indoors! A Campaign for Safer Birds and Cats at http://www.abcbirds.org/cats/catsindoors.htm.

8. The American Kennel Club compiled registration figures for 150 breeds of dogs in 2001. The top five were Labrador retriever, golden retriever, German shepherd, dachshund, and beagle. The Cat Fanciers' Association recognized forty breeds in 2001, the top five of which were Persian, Maine coon, Siamese, Abyssinian, and exotic.
9. The term "breed" refers to a division of a species with separate, distinctive characteristics that can be predicted genetically and reproduced consistently. "Breed" is not a term used in the science of taxonomy, which classifies organisms according to *kingdom, phylum, class, order, family, genus, species*, and *subspecies*. The domestic cat belongs to the cat family, Felidae, which has various species divided into three genera: *Panthera, Acinonyx*, and *Felis*. The domestic cat is *Felis catus*, and the many breeds known today (Siamese, Manx, American shorthair, for example) are variations of that species. The dog family, Canidae, with its thirty-eight species, is more diverse.
10. For those who are unfamiliar with cats, I refer here to hair balls, which cats occasionally vomit in the least opportune places.
11. For this and other examples, see Serpell 1986, chap. 2.
12. There are at least two caveats about such studies, as well as about their meta-analysis. One is that the categories of "pet owners" and "non-owners" do not consider previous ownership. People who were "non-owners" at the time of the study may once have had pets; studies suggest that most people (from 88 to 94 percent) have had pets at some time during their lives (see Kidd and Kidd 1980; Podberscek and Gosling 2000, 161). A second difficulty

is that differences in methodology, variables tested, and psychometric instruments used make direct comparisons of most of the owner–non-owner studies impossible.

13. For a review, see Podberscek and Gosling 2000.
14. Gardner (1980) explores civil inattention and violation, explaining that children serve the same function as dogs.
15. See Robins et al. 1991 and West 1999 for similar accounts.
16. The current *Catechism of the Catholic Church* (1994, 580–81) reminds the faithful that it is "unworthy to spend money on [animals] that should as a priority go to the relief of human misery."
17. In addition to the scholarly references about our obligations to animals, I want to mention that the fox told the little prince, "You become responsible for what you have tamed" (Saint-Exupéry 1971 [1943], 64).
18. Joanna Swabe (2000) offers a thorough discussion of the ambivalence of veterinarians toward client-requested mutilation.
19. In addition, the "natural" connection between children and animals is questionable. Aline Kidd and Robert Kidd (1987) found that it is normal for children to show no interest in animals.

CHAPTER 2

1. For a discussion of the psychological mechanisms that allow us to distinguish between animals we love and those we eat, see Plous 1993.
2. Some biblical scholars argue that interpretations of the word that justify using animals as we please misrepresent the original Hebrew. A rival interpretation translates the original word as "stewardship," a weak form of anthropocentrism indicating a "God-given responsibility to *care for* the earth" (Linzey 1998, 287; emphasis added) instead of the right to *rule over* it (see also Cohen 1989).
3. A compelling and comprehensive discussion of dominionism and its consequences appears in Scully 2002.
4. Turning Darwin on his head, Plato held that humans existed before animals (see Sorabji 1993, 9–12).
5. Other qualities that have allegedly distinguished humans from, and that make them superior to, animals include speech, physical beauty, religion, private property, and tool use. Jane Goodall (1990) debunked the last of these when she observed the chimpanzee, David Greybeard, not only using a tool but also making one.
6. It is worth mentioning that the era produced not only Thomas Aquinas but also St. Francis of Assisi, known for preaching from a humane perspective and for having an uncanny ability to attract birds and wild animals (see Arm-

strong 1973). There are also earlier pro-animal strands in the Christian tradition, notably John Chrysostom of the fourth century.

7. For a thorough treatment of the indirect duty view, see Niven 1967, 29–37. For the flaws in the argument, see DeGrazia 1996.
8. Aquinas endorsed the Aristotelian astronomy and physics that placed the Earth at the center of a finite universe. The fates of Copernicus and Galileo illustrate what lay in store for anyone who opposed this model.
9. See Revelation 22:15.
10. For a thorough treatment of the history of hunting, as well as technology, ethics, and more, see the entry in Bekoff 1998.
11. The investment of the dog with nearly human status cuts both ways. Mary Douglas (1966) makes the point that the liminal position of the dog on the border between human and non-human constituted sufficient reason for regarding it as potentially unclean. For a review of literature addressing this among various cultures, see Serpell 1995.
12. This was identified by Kenneth Clark (1977).
13. According to Serpell 1986, witchcraft was the common accusation in Britain, whereas on the continent the charge was chiefly bestiality.
14. Cats remain easy targets for torture today. The night before I wrote this section, the evening news reported that a beheaded cat had been found in a local park, the second in recent months.
15. Bentham, considered the founder of utilitarianism, argued that the capacity to suffer pain was accompanied by and necessary for a similar capacity to feel pleasure. This "governance of two sovereign masters," as he called it—of pain and pleasure—was the core of his utilitarian doctrine, outlined in *The Principles of Morals and Legislation,* from which the famous "Can they suffer?" quote comes. Defining utility as the ability of some object or activity to produce pleasure or prevent pain, he argued that animals as well as humans had utilitarian interests. For a concise discussion of the pitfalls of Bentham's position, see Francione 2000, chap. 6.
16. In the German laws, in contrast to Martin's Act, prosecution required that the cruelty had occurred publicly and had offended human observers. This made the German laws an extension of the indirect duty view (see Maehle 1994).
17. The idea also spawned a third interpretation that is beyond the scope of this discussion: outright rejection, especially within fundamentalist movements.
18. Another justification of hunting, and one in use today, claims that hunters help the evolutionary process by culling animals who otherwise would starve because of overpopulation. However, hunters encourage overpopulation and reverse evolution. They kill the biggest and healthiest animals, whereas natural predators—that is, non-humans—kill the weakest. They select the sex

of their prey, thereby interfering with reproduction rates. For more on this and other hunting narratives, see Kheel 1995 (see also Einwhoner 1999).

19. The twentieth-century version of the Romantic attitude produced animal-rights movements. In animal protection (or welfare), the goal is to minimize harm. In animal-rights movements, the goal is an end to all animal exploitation—in zoos and wildlife-park exhibits, as research subjects and food sources, and as pets (see Bekoff 1998; Franklin 1999, 27–33).
20. The terms "humane society" and "SPCA" may be used by any organization. No national agency oversees shelters or societies for the prevention of cruelty to animals. The Humane Society of the United States and American Humane Association offer guidelines and policy recommendations that many shelters follow, but they have no obligation to do so. For thorough histories of the humane movement, see Coleman 1924 and Niven 1967.
21. The name "pound" comes from the word "impound." There are three kinds of shelters in the United States: 1) Municipal animal-control facilities run by city or township governments. Most of these are "pounds," or impound facilities that hold animals picked up or turned in for a given period of time, after which the animals are euthanized or, in some cases, transferred to other shelters; 2) private, nonprofit facilities governed by boards of directors; and 3) private nonprofits that also have government contracts to provide animal control.
22. In the United States, unclaimed animals were routinely sold for use in research until 1979 (see Finsen and Finsen 1994, 61). The practice, called "pound seizure," was required by the Metcalf-Hatch Act. New York, Minnesota, and several other states, as well as municipalities, had pro–pound-seizure laws. Individual states began to repeal pound-seizure laws in the 1970s, but the turning point was the repeal of Metcalf-Hatch in 1979.
23. White was also an abolitionist and a Quaker, which provides support for the "extension thesis." For a biographical sketch, see Unti 1998.
24. In 1897, the Women's Branch became the Women's Pennsylvania Society for the Prevention of Cruelty to Animals. It was incorporated independently to manage properly any revenue the Women's Branch might receive, and White and several other women secured the charter to do so.
25. "Baiting" involves tying the animal to a stake and allowing other animals—often dogs, and several at once—to attack it. Animals that are customarily baited include badgers, mules, horses, bears, and apes, in addition to bulls. These "sports" were popular even among English royalty until the seventeenth century, when the upper classes began to have less tolerance for cruel animal sports. Considerable debate surrounds the reasons for this change. For one view, see Elias and Dunning 1986; Tester 1992; and Elias 1994. For another, see Franklin 1999. Britain outlawed bull baiting in 1835, and cockfighting became illegal in 1849, but the elite sports of foxhunting and sport-

fishing continue (see also Franklin 1999, 22–24). The United States did not begin to outlaw dogfighting until after the Civil War, and, at the time of this writing, cockfighting remains legal in three states.

26. The targeting of the working class is an important difference between the anti-cruelty movement and the antivivisectionist movement, the latter of which targeted the educated elite (see Sperling 1988, 32).
27. Similarly, Glen Elder, Jennifer Wolch, and Jody Emel (1998) outline the ways in which animal practices in the late twentieth century function to racialize certain groups.
28. Other examples include the popularity of aquariums, house plants, and caged birds.
29. William Danforth, using a borrowed $12,000, established a feed store in St. Louis, Missouri. In the early 1900s, he began selling a whole-wheat cereal for humans under the name Purina, from the company slogan, "Where purity is paramount." A well-known health advocate named Dr. Ralston later endorsed the cereal. and the company was officially renamed Ralston Purina in 1902. The checkerboard logo comes from Danforth's childhood recollections of the cloth from which his mother made clothing for his siblings and him.
30. During World War I, noting the enthusiasm with which soldiers responded to the word "chow" for food, the company changed the name of what were known as "feeds" to "chow." Purina Dog Chow and Cat Chow are still sold.
31. Cat litter was created in 1948. In January of that year, a Michigan woman named Kay Draper, who, like everyone else who had cats at the time, used sand in the litter box went to the local sand pile but found it frozen. She tried ashes as a replacement, but the cat left sooty paw prints all over the house. Thinking that she would try sawdust, she went to the Lowe family's coal, ice, and sawdust company. Ed Lowe had a mound of dried, granulated, mineral clay that he had tried to persuade chicken farmers to use as nesting material. Draper agreed to try some in her cat's litter pan. It worked so well that she returned for more and told her friends about it. Lowe began filling paper bags with it and marking them "Kitty Litter." Thus began a 2.5 billion pound, $708 million a year industry with its own lobby in Washington, D.C. One in three retail dollars spent on pet supplies goes toward cat litter. Although the "fresh scent," "premium," "clumping," and "scoopable" litters of today bear little resemblance to Lowe's product, the primary ingredient is still Fuller's Earth, a crystalline clay mineral that is also the main ingredient in the diarrhea treatment Kaopectate. Lowe's invention "helped lay the groundwork for the cat's eventual emergence as the most plentiful four-legged pet in America. Kitty litter, in the words of one industry official, has done for the cat what air conditioning did for Houston: It took the worry out of being close" (Maggitti 1996, 48).

32. For example, the local government in Boulder, Colorado, established a licensing law in 1871 in response to citizens' fears of and complaints about roaming packs of dogs. Constables had orders to kill non-licensed dogs on sight, and each dead dog brought a bounty of $1.

CHAPTER 3

1. In agility, a dog and a handler run an obstacle course, with the handler directing the dog while the dog navigates the obstacles, which call on the dog's speed, balance, jumping, and other abilities, as well as communication between dog and handler. For more information, go to http://www.dogpatch.org/agility or http://www.usdaa.com. Flyball is a fast-paced sport that is especially good for high-energy, ball-crazed dogs. For more information, go to http://www.flyballdogs.com or http://www.flyball.org.
2. The notion of shaping behavior originated in marine mammal training in the 1960s. The pioneer in the field was Karen Pryor, a dolphin trainer who extended the approach to the training of other animals (see Pryor 1986; see also Donaldson 1996; Owens and Eckroate 1999). Granted, people had used nonviolent approaches before the recent boom in positive training—notably, Montague Stevens (1990 [1943]), who at the turn of the century was considered "an eccentric—at best" (Derr 1997, 326). Fortunately, nonviolence is no longer considered unconventional, and gentle trainers have their own associations, websites, workshops, and seminars (see http://www.apdt.com). The *Guide to Humane Dog Training*, a guide for guardians, is available from the American Humane Association (http://www.americanhumane.org) and *Professional Standards for Dog Trainers* can be obtained from the Delta Society (http://www.deltasociety.org).
3. For more on"Disneyfication," see Milekic 1998; see also Lawrence 1986.
4. Donald Griffin is particularly well known for his research demonstrating that bats navigate and hunt using echolocation.
5. More thorough treatments of cognitive ethology appear in Ristau 1991 and Allen and Bekoff 1997.
6. For thorough discussions of anthropomorphism, see Mitchell et al. 1997 and Crist 1999.
7. Indeed, some have even dismissed Goodall's empathy as "rather anthropomorphic" (Jasper and Nelkin 1992, 199, n. 10) because she describes chimps as her friends. To me, the possibility of friendship with members of a species with whom we share more than 98 percent of our DNA does not seem outrageous.
8. Here I must beg the indulgence of my family, who, recalling my nearly disastrous attempt to learn to play tennis, will surely question my use of this example. I emphasize that it is Shapiro's, not mine.

CHAPTER 4

1. Gagnon's argument has to do with the material changes that influenced mental life, and consequently the self, during the nineteenth century, when shop windows and department stores were invented. Gagnon also examines rail travel, photography, and the rise of literacy and mass-produced reading material as influences that increased the number of "internal conversations" that occur in the self.
2. Berger ultimately draws very different conclusions about animals from those I offer in this book.

CHAPTER 5

1. Alger and Alger 2003, chap. 6, found five primary reasons that people chose a particular cat for adoption.
2. Anthropologists say that human beings are a neotenous species because we retain into adulthood features that were originally juvenile in our primate ancestors. Our flat faces, domed skulls, hairless bodies, small teeth, and large eyes appear in infant and juvenile apes. For further discussion, see Lawrence 1986.
3. Posage et al. 1998 found that black coat color among dogs was strongly associated with euthanasia at shelters that use that practice for space. A contributing factor may have been that most of the black dogs were large, which discourages many potential adopters.
4. I borrow these categories from Csikszentmihalyi and Robinson 1990.
5. See Aronson 1999, chap. 8, for a review of these studies.
6. Ann Landers once warned readers: "Don't accept your dog's admiration as conclusive evidence that you are wonderful."
7. This tendency to like animals who seem to like us probably explains the popularity of golden retrievers and Labrador retrievers, which are bred to be sociable family dogs.
8. To teach a dog to make eye contact, a volunteer trainer holds a treat near his or her eyes. When the dog looks at the treat, he or she gets the reward. Gradually, the trainer introduces the phrase, "Look at me!" and begins to move the treat away from the eyes. The dog then gets the treat when he or she retains eye contact, focusing attention on the person instead of the treat. In short, a dog's willingness to make eye contact is so important that The Shelter makes a concerted effort to bring dogs to that point. In addition, people place so much value on dogs' eye contact that any truly adoptable dog must be able to engage in it.
9. See Hochschild 1983, app. A, for a discussion of the various models of emotions.

10. Although not essential for the present discussion, I can offer a brief answer to the question of why this occurred. In part, the fading of intensity accompanied the demands of an advanced industrial society, combined with new family patterns and increasing consumerism. Peter Stearns (1994) adds the importance of a twentieth-century shift from excessive religious spirituality and an increasing emphasis on health.
11. Americans, more than members of many other cultures, tend to define the emotions they find unpleasant as dangerous and not potentially useful. The Chinese, for example, find certain emotions such as guilt and jealousy difficult or unpleasant but also recognize that these feelings could have important functions. Americans differ strikingly from several other cultures not only in their disapproval of certain emotions but also in their desire to conceal those emotions and their pride in the ability to do so. Consequently, the American emotional vocabulary manifests a relatively simplistic pleasure–pain dichotomy. For a thorough discussion, see Sommers 1984.

CHAPTER 6

1. Myers 1998 and Sanders 1999 offer thorough discussions of the Aristotelian and neo-Cartesian dimensions of Mead's thought.
2. It amazes me that Mead supposedly had a bulldog who accompanied him everywhere.
3. For more examples of infancy studies with non-human animals, see Myers 1998, especially chap. 4.

CHAPTER 7

1. "Agency" most often appears paired with its putative antithesis in a controversy referred to as the "structure–agency debate." This is an ongoing argument over whether sociological (external) or psychological (internal) factors are the "correct" ones to use to explain human action. Sociologists usually emphasize structural explanations, leaving agency (and culture) "as conceptual underdogs" (Hays 1994, 58). For a discussion of the history and a potential resolution of the debate, see Emirbayer and Mische 1998 (see also Rubenstein 2001; Côté and Levine 2002). Further insights into some of these various uses of the word "agency" appear in Davies 2000, chap. 4.
2. For a discussion of when agency is present but not conscious, and when it involves consciousness, see Dawkins 1998.
3. For more on motor plans, see Stern 1985, particularly the example of handwriting on page 78. When subjects were asked to write their signatures twice—once on paper in normal size and again on a blackboard much larger than normal—the two signatures were alike once they were aligned

in size. Although the actions involved two different muscle groups, the resulting signatures were the same. This happens because the motor plan for the signature "lives" in the mind and can be transferred to different sets of muscles.

4. The Algers reveal that some cat guardians clearly disliked the idea of training cats. For example, one respondent is quoted as saying, "If I were going to train something, I would get a dog" (Alger and Alger 2003, 21).
5. Although animals do not name one another, there is evidence that some species recognize other individuals' "signature" calls or whistles: See Masson and McCarthy 1995, 36–37.
6. For a perspective on how cats know this, as well as how animals seem to know when other things are going to happen, see the work of Rupert Sheldrake (1999). Sheldrake and Bekoff 2000 discusses Sheldrake's research and address science's prejudices about what animals allegedly can and cannot do.
7. Contrary to Darwin, Masson and McCarthy 1997, 13, point out that "not all actions driven by emotion have survival value." They offer the examples of animals who put themselves at physical risk to mourn a dead loved one or those who adopt orphaned babies, thus not passing on their own genes.
8. Stern provides the example of puppets, who "have little or no capacity to express categories of affect by way of facial signals, and their repertoire of conventionalized gestural or postural affect signals is usually impoverished. *It is from the way they move in general that we infer the different vitality affects from the activation contours they trace.* Most often, the characters of different puppets are largely defined in terms of particular vitality affects; one may be lethargic, with drooping limbs and hanging head, another forceful, and still another jaunty" (Stern 1985, 56; emphasis added).
9. A convincing case for pre-verbal memory can be made among victims of early childhood abuse. Although the child lacks words to describe what occurred, and perhaps cannot even recall distinctly, memories nevertheless endure. They can be triggered by, say, the sound of a belt being removed or by a particular odor, such as cigarette smoke or alcohol.
10. I believe Mark Twain said that if two people agree on everything, one of them is redundant.

CHAPTER 8

1. The ferret's lethargy, reference to which will puzzle those who have interacted with this energetic species, came from eating a high-calorie meal.
2. For additional discussion, see Irvine 2001.
3. Researchers have distinguished several different types of play, and I discuss two here. I examine social object play, which refers to two partners playing with a toy (or an object that serves as one). When guardians played with

their animals, they most often engaged in social object play, as in when they threw a ball for a dog or dangled a string for a cat. I also discuss social play, which is when two partners play together. See Bekoff and Byers 1981 for further discussion.

4. This is roughly consistent with the findings of Reinhold Berghler's detailed study of nearly 300 cat guardians (Berghler 1989).
5. To be clear, Bekoff and Byers's definition of play does not include intentionality. The play bow signals intention to play, but the intentional elements of play itself depend on the investigator's concept of intention. This is the topic of a long philosophical debate reviewed in Allen and Bekoff 1997, 93, who write that "ultimately, it might be found that play is an intentional activity; however, to include this in the definition of play would be premature. . . . The relevance of intentionality to play is a matter for empirical investigation not *a priori* definition, and we urge its investigation as such."
6. When I first heard about Clever Hans, it struck me that people dismissed him for being able to do what many humans seem incapable of doing.
7. Here, I am indebted to John Hewitt's (2000) symbolic-interactionist social psychology.

CONCLUSION

1. The data on households comes from American Veterinary Medical Association surveys; the 90 percent comes from a Gallup poll (see Gallup 1996).
2. Although I will address only companion animals here, this extends to all species and entails decisions about meat eating, experimentation, hunting, wearing fur and leather, and various forms of entertainment. The literature that can assist in developing an ethical stance includes Regan 1983; Singer 1990; DeGrazia 1996; Francione 2000; and Wise 2000.
3. Further discussion appears in Midgley 1983; Rollin 1992; and Bekoff 1998.
4. For a thorough treatment of basic rights, see Shue 1996.
5. Tethering a dog in a public place is illegal in many places because most dog bites occur when dogs are tied up. Children will often approach strange dogs in this way, only to be bitten on the face by a dog whose ability to escape the potentially threatening situation is limited.

References

Adler, Patricia A., and Peter Adler. 1999. "Transience and the Postmodern Self: The Geographic Mobility of Resort Workers." *Sociological Quarterly* 40: 31–58.

Alger, Janet M., and Steven F. Alger. 1997. "Beyond Mead: Symbolic Interaction between Humans and Felines." *Society & Animals* 5: 65–81.

———. 1999. "Cat Culture, Human Culture: An Ethnographic Study of a Cat Shelter." *Society & Animals* 7: 199–218.

———. 2003. *Cat Culture: The Social World of a Cat Shelter.* Philadelphia: Temple University Press.

Allen, Colin, and Marc Bekoff. 1997. *Species of Mind: The Philosophy and Biology of Cognitive Ethology.* Cambridge, Mass.: MIT Press.

American Veterinary Medical Association. 2002. *U.S. Pet Ownership and Demographics Sourcebook.* Schaumburg, Ill.: Center for Information Management of the American Veterinary Medical Association.

Apter, Michael J. 1991. "A Structural Phenomenology of Play." Pp. 13–29 in *Adult Play: A Reversal Theory Approach,* ed. John H. Kerr and Michael J. Apter. Amsterdam: Swets and Zeitlinger.

Arluke, Arnold. 1991. "Going into the Closet with Science: Information Control among Animal Experimenters." *Journal of Contemporary Ethnography* 20: 306–30.

———. 1994. "'We Build a Better Beagle': Fantastic Creatures in Lab Animal Ads." *Qualitative Sociology* 17: 143–58.

Arluke, Arnold, and Clinton R. Sanders. 1996. *Regarding Animals*. Philadelphia: Temple University Press.

Arluke, Arnold, and Boria Sax. 1992. "Understanding Nazi Animal Protection and the Holocaust." *Anthrozoös* 5: 176–91.

Arluke, Arnold, Randy Frost, Gail Steketee, Gary Patronek, Carter Luke, Edward Messner, Jane Nathanson, and Michelle Papazian. 2002. "Press Reports of Animal Hoarding." *Society & Animals* 10: 113–35.

Armstrong, Edward Allworthy. 1973. *Saint Francis: Nature Mystic*. Berkeley: University of California Press.

Arnheim, Rudolph. 1971. *Entropy and Art*. Berkeley: University of California Press.

———. 1982. *The Power of the Center*. Berkeley: University of California Press.

Aronson, Elliot. 1999. *The Social Animal*, 8th ed. New York: W. H. Freeman.

Bateson, Patrick, and Dennis C. Turner. 1988. "Questions about Cats." Pp. 193–201 in *The Domestic Cat: The Biology of Its Behaviour*, ed. Dennis C. Turner and Patrick Bateson. Cambridge: Cambridge University Press.

Beck, Alan, and Aaron Katcher. 1996. *Between Pets and People: The Importance of Animal Companionship*, rev. ed. West Lafayette, Ind.: Purdue University Press.

Beck, Alan M., Gail F. Melson, Patricia L. da Costa, and Ting Liu. 2001. "The Educational Benefits of a Ten-Week Home-Based Wild Bird Feeding Program for Children." *Anthrozoös* 14: 19–28.

Becker, Gary S. 1975. *Human Capital*, 2nd ed. New York: National Bureau of Economic Research; distributed by Columbia University Press.

Bekoff, Marc. 1977. "Social communication in canids: Evidence for the evolution of a stereotyped mammalian display." *Science* 197:1097–99.

———. 1995. "Play Signals as Punctuation: The Structure of Social Play in Canids." *Behaviour* 132:419–29.

———. 2000. *Strolling with Our Kin: Speaking for and Respecting Voiceless Animals*. New York: Lantern/Booklight.

———. 2002. *Minding Animals: Awareness, Emotions, and Heart*. Oxford: Oxford University Press.

Bekoff, Marc, ed. 1998. *Encyclopedia of Animal Rights and Animal Welfare*. Westport, Conn.: Greenwood.

Bekoff, Marc, and Colin Allen. 1998. "Intentional Communication and Social Play." Pp. 97–114 in *Animal Play: Evolutionary, Comparative, and Ecological Perspectives*, ed. Marc Bekoff and John Byers. Cambridge: Cambridge University Press.

Bekoff, Marc, and John Byers. 1981. "A Critical Reanalysis of the Ontogeny of Mammalian Social and Locomotor Play: An Ethological Hornet's Nest." In

Behavioral Development, ed. Klaus Immelmann, George W. Barlow, Lewis Petrinovich, and Mary Main. Cambridge: Cambridge University Press.

Bentham, Jeremy. 1988 (1781). *The Principles of Morals and Legislation*. Amherst, N.Y.: Prometheus.

Berger, John. 1980. *About Looking*. New York: Pantheon/Random House.

Berger, Peter, and Thomas Luckmann. 1967. *The Social Construction of Reality: A Treatise in the Sociology of Knowledge*. Garden City, N.Y.: Doubleday Anchor.

Berghler, Reinhold. 1989. *Man and Cat: The Benefits of Cat Ownership*. Oxford: Blackwell Scientific.

Birke, Lynda. 1994. *Feminism, Animals and Science*. Buckingham, U.K.: Open University Press.

Bogdan, Robert, and Steven Taylor. 1989. "Relationships with Severely Disabled People: The Social Construction of Humanness." *Social Problems* 36: 135–48.

Bond, Simon. 1981. *A Hundred and One Uses for a Dead Cat*. London: Methuen.

Bourdieu, Pierre. 1986. "The Forms of Capital." Pp. 241–58 in *Handbook of Theory and Research for the Sociology of Education*, ed. John G. Richardson. New York: Greenwood.

Brazelton, T. Berry. 1984. "Four Stages in the Development of Mother–Infant Interaction." Pp. 19–34 in *The Growing Child in Family and Society*, ed. Noboru Kobayashi and T. Berry Brazelton. Tokyo: University of Tokyo Press.

Brestrup, Craig. 1997. *Disposable Animals: Ending the Tragedy of Throwaway Pets*. Leander, Tex.: Camino Bay Books.

Bruner, Jerome, and David A. Kalmar. 1998. "Narrative and Metanarrative in the Construction of Self." Pp. 308–31 in *Self-Awareness: Its Nature and Development*, ed. Michel Ferrari and Robert J. Sternberg. New York: Guilford.

Budiansky, Stephen. 1992. *The Covenant of the Wild: Why Animals Chose Domestication*. New Haven, Conn.: Yale University Press.

———. 2002. "The Character of Cats." *Atlantic Monthly*, vol. 289 (June), 75–77.

Burghardt, Gordon M. 1998. "The Evolutionary Origins of Play Revisited: Lessons from Turtles." Pp. 1–26 in *Animal Play: Evolutionary, Comparative, and Ecological Perspectives*, ed. Marc Bekoff and John Byers. Cambridge: Cambridge University Press.

Byrne, Donn. 1969. "Attitudes and Attraction." Pp. 36–89 in *Advances in Experimental Social Psychology*, vol. 4, ed. Leonard Berkowitz. New York: Academic Press.

Carson, Rachel. 1962. *Silent Spring*. Boston: Houghton Mifflin.

Cartmill, Matt. 1997. "History of Ideas Surrounding Hunting." Pp. 197–99 in *Encyclopedia of Animal Rights and Animal Welfare*, ed. Marc Bekoff. Westport, Conn.: Greenwood.

Catechism of the Catholic Church. 1994. Mahwah, N.J.: Paulist Press.
Clark, Kenneth. 1977. *Animals and Men: Their Relationship as Reflected in Western Art from Prehistory to the Present Day*. New York: Morrow.
Clark, Stephen R. L. 1982. *The Nature of the Beast: Are Animals Moral?* Oxford: Oxford University Press.
Clutton-Brock, Juliet. 1981. *Domesticated Animals from Early Times*. London: Heinemann.
———. 1994. "The Unnatural World: Behavioural Aspects of Humans and Animals in the Process of Domestication." Pp. 23–35 in *Animals and Human Society: Changing Perspectives*, ed. Aubrey Manning and James Serpell. London: Routledge.
———. 1995. "Origins of the Dog: Domestication and Early History." Pp. 8–20 in *The Domestic Dog: Its Evolution, Behaviour and Interactions with People*, ed. James Serpell. Cambridge: Cambridge University Press.
Cohen, Jeremy. 1989. *"Be Fertile and Increase, Fill the Earth and Master It": The Ancient and Medieval Career of a Biblical Text*. Ithaca, N.Y.: Cornell University Press.
Coleman, Sidney H. 1924. *Humane Society Leaders in America*. Albany, N.Y.: American Humane Association.
Collis, Glyn M., and June McNicholas. 1998. "A Theoretical Basis for Health Benefits of Pet Ownership: Attachment versus Psychological Support." Pp. 105–22 in *Companion Animals in Human Health*, ed. Cindy C. Wilson and Dennis C. Turner. Thousand Oaks, Calif.: Sage.
Coppinger, Raymond, and Lorna Coppinger. 2001. *Dogs: A Startling New Understanding of Canine Origin, Behavior, and Evolution*. New York: Scribner.
Coppinger, Raymond, and Richard Schneider. 1995. "Evolution of Working Dogs." Pp. 22–47 in *The Domestic Dog: Its Evolution, Behaviour and Interactions with People*, ed. James Serpell. Cambridge: Cambridge University Press.
Côté, James E., and Charles G. Levine. 2002. *Identity Formation, Agency, and Culture: A Social Pscyhological Synthesis*. Mahwah, N.J.: Lawrence Erlbaum.
Crandall, Lee S. 1917. *Pets: Their History and Care*. New York: Henry Holt.
Crist, Eileen. 1999. *Images of Animals: Anthropomorphism and Animal Mind*. Philadelphia: Temple University Press.
Crooks, Kevin R., and Michael E. Soulé. 1999. "Mesopredator Release and Avifaunal Extinctions in a Fragmented System." *Nature* 400: 563–66.
Csikszentmihalyi, Mihaly. 1990. *Flow: The Psychology of Optimal Experience*. New York: Harper and Row.
———. 1997. *Finding Flow: The Psychology of Engagement with Everyday Life*. New York: Basic Books.
Csikszentmihalyi, Mihaly, and Rick E. Robinson. 1990. *The Art of Seeing: An Interpretation of the Aesthetic Encounter*. Malibu, Calif.: J. Paul Getty Trust.

Damasio. Antonio. 1999. *The Feeling of What Happens: Body and Emotion in the Making of Consciousness*. New York: Harcourt Brace.

Darnton, Robert. 1985. *The Great Cat Massacre and Other Episodes in French Cultural History*. New York: Penguin.

Darwin, Charles. 1859. *On the Origin of Species*. London: John Murray.

———. 1871 (1936). *The Descent of Man and Selection in Relation to Sex*. New York: Random House.

———. 1872 (1965). *The Expression of the Emotions in Man and Animals*. Chicago: University of Chicago Press.

Davies, Bronwyn. 2000. *A Body of Writing 1990–1999*. Walnut Creek, Calif.: AltaMira.

Dawkins, Marian Stamp. 1998. *Through Our Eyes Only? The Search for Animal Consciousness*. Oxford: Oxford University Press.

DeGrazia, David. 1996. *Taking Animals Seriously: Mental Life and Moral Status*. Cambridge: Cambridge University Press.

Derr, Mark. 1997. *Dog's Best Friend: Annals of the Dog–Human Relationship*. New York: Henry Holt.

de Swaan, Abram. 1981. "The Politics of Agoraphobia: On Changes in Emotional and Relational Management." *Theory and Society* 10: 359–85.

Dewey, John. 1934. *Art as Experience*. New York: Perigree.

Diffey, T. J. 1986. "The Idea of Aesthetic Experience." Pp. 3–12 in *Possibility of the Aesthetic Experience*, ed. Michael H. Mitias. Dordrecht, The Netherlands: Martinus Nijhoff.

Dion, Karen, Ellen Berscheid, and Elaine Walster (Hatfield). 1972. "What Is Beautiful Is Good." *Journal of Personality and Social Psychology* 24: 285–90.

Donaldson, Jean. 1996. *The Culture Clash*. Berkeley, Calif.: James and Kenneth Publishers.

Douglas, Mary. 1966. *Purity and Danger: An Analysis of the Concepts of Pollution and Taboo*. New York: Routledge and Kegan Paul.

Dowd, James J. 1991. "Social Psychology in a Postmodern Age: A Discipline without a Subject." *American Sociologist* 22:188–209.

Einwhoner, Rachel L. 1999. "Practices, Opportunity, and Protest Effectiveness: Illustrations from Four Animal Rights Campaigns." *Social Problems* 46: 169–86.

Elder, Glen Jennifer Wolch, and Jody Emel. 1998. "*Le Pratique Sauvage*: Race, Place, and the Human–Animal Divide." Pp. 72–90 in *Animal Geographies: Place, Politics, and Identity in the Nature–Culture Borderlands*, ed. Jennifer Wolch and Jody Emel. London: Verso.

Elias, Norbert. 1994. *The Civilizing Process*. Oxford: Blackwell.

Elias, Norbert, and Eric Dunning. 1986. *Quest for Excitement*. Oxford: Blackwell.

Eliot, George. 1880. *Scenes of Clerical Life*. New York: William L. Allison.

Emel, Jody. 1998. "Are You Man Enough, Big and Bad Enough? Wolf Eradication in the U.S." Pp. 91–116 in *Animal Geographies: Place, Politics, and Identity in the Nature–Culture Borderlands*, ed. Jennifer Wolch and Jody Emel. London: Verso.

Emel, Jody, and Jennifer Wolch. 1998. "Witnessing the Animal Moment." Pp. 1–24 in *Animal Geographies: Place, Politics, and Identity in the Nature–Culture Borderlands*, ed. Jennifer Wolch and Jody Emel. London: Verso.

Emirbayer, Mustafa, and Ann Mische, 1998. "What Is Agency?" *American Journal of Sociology* 103: 962–1023.

Feingold, A. 1990. "Gender Differences in Effects of Physical Attractiveness on Romantic Attraction: A Comparison across five Research Paradigms." *Journal of Personality and Social Psychology* 59: 981–93.

Finsen, Lawrence, and Susan Finsen. 1994. *The Animal Rights Movement in America: From Compassion to Respect*. New York: Twayne.

Fisher, John Andrew. 1991. "Disambiguating Anthropomorphism: An Interdisciplinary Review." Pp. 49–85 in *Perspectives in Ethology, Vol. 9: Human Understanding and Animal Awareness*, ed. P. Bateson and P. Klopfer. New York: Plenum.

Flynn, Clifton P. 1999. "Animal Abuse in Childhood and Later Support for Interpersonal Violence in Families." *Society & Animals* 7: 161–72.

———. 2000a. "Woman's Best Friend: Pet Abuse and the Role of Companion Animals in the Lives of Battered Women." *Violence Against Women* 6: 162–77.

———. 2000b. "Battered Women and Their Animal Companions: Symbolic Interaction between Human and Nonhuman Animals." *Society & Animals* 8: 99–127.

Fogle, Bruce, ed. 1981. *Interrelations between People and Pets*. Springfield, Ill.: Charles C. Thomas.

Forster, E. M. 1921. *Howard's End*. New York: Alfred A. Knopf.

Francione, Gary L. 1995. *Animals, Property, and the Law*. Philadelphia: Temple University Press.

———. 1996. *Rain Without Thunder: The Ideology of the Animal Rights Movement*. Philadelphia: Temple University Press.

———. 2000. *Introduction to Animal Rights: Your Child or the Dog?* Philadelphia: Temple University Press.

Franklin, Adrian. 1999. *Animals and Modern Cultures: A Sociology of Human–Animal Relations in Modernity*. London: Sage.

Franklin, Adrian, Bruce Tranter, and Robter White. 2001. "Explaining Support for Animal Rights: A Comparison of Two Recent Approaches to Human, Nonhuman Animals, and Postmodernity." *Society & Animals* 9: 127–44.

Gagnon, John H. 1992. "The Self, Its Voices, and Their Discord." Pp. 221–43 in *Investigating Subjectivity*, ed. Carolyn Ellis and Michael Flaherty. Newbury Park, Calif.: Sage.

Gallup, Alec. 1996. "Gallup Poll: Dog and Cat Owners See Pets as Part of Family." *Minneapolis Star Tribune,* October 28, E10.

Gardner, Carol Brooks. 1980. "Passing By: Street Remarks, Address Rights, and the Urban Female." *Sociological Inquiry* 50: 328–56.

Geertz, Clifford. 1984. "'From the Native's Point of View': On the Nature of Anthropological Understanding." Pp. 123–37 in *Culture Theory,* ed. Richard Shweder and Rober LeVine. Cambridge: Cambridge University Press.

Gergen, Kenneth. 1991. *The Saturated Self: Dilemmas of Identity in Contemporary Life.* New York: Basic Books.

Gerhards, Jürgen. 1989. "The Changing Culture of Emotions in Modern Society." *Social Science Information* 28: 737–54.

Giddens, Anthony. 1991. *Modernity and Self Identity.* Cambridge: Polity.

Goffman, Erving. 1959. *The Presentation of Self in Everyday Life.* Garden City, N.Y.: Anchor Books.

———. 1963. *Behavior in Public Places.* New York: Free Press.

———. 1967. *Interaction Ritual.* Garden City, N.Y.: Anchor Books.

———. 1974. *Frame Analysis: An Essay on the Organization of Experience.* Cambridge, Mass.: Harvard University Press.

Gombrich, Ernest H. 1960. *Art and Illusion: A Study in the Psychology of Pictorial Representation.* Princeton, N.J.: Princeton University Press.

———. 1979. *Ideals and Idols: Essays on Values in History and in Art.* Oxford: Phaidon.

Goodall, Jane. 1990. *Through a Window: My Thirty Years with the Chimpanzees of Gombe.* Boston: Houghton Mifflin.

———. 1999. *Reason for Hope: A Spiritual Journey.* New York: Warner Books.

Gordon, Steven L. 1981. "The Sociology of Sentiments and Emotions." Pp. 562–92 in *Social Psychology: Sociological Perspectives,* ed. Morris Rosenberg and Ralph H. Turner. New York: Basic Books.

Griffiths, Huw, Ingrid Poulter, and David Sibley. 2000. "Feral Cats in the City." Pp. 56–70 in *Animal Spaces, Beastly Places: New Geographies of Human–Animal Relations,* ed. Chris Philo and Chris Wilbert. London: Routledge.

Griffin, Donald R. 1976. *The Question of Animal Awareness: Evolutionary Continuity of Mental Experience.* New York: Rockefeller University Press.

———. 1992. *Animal Minds.* Chicago: University of Chicago Press.

Gubrium, Jaber. 1986. "The Social Preservation of Mind: The Alzheimer's Disease Experience." *Symbolic Interaction* 6: 37–51.

Guttman, Giselher. 1981. "The Psychological Determinants of Keeping Pets." Pp. 89–98 in *Interrelations between People and Pets,* ed. Bruce Fogle. Springfield, Ill.: Charles C. Thomas.

Hall, Libby. 2000. *Prince and Other Dogs: 1850–1940.* New York: Bloomsbury.

Halle, David. 1993. *Inside Culture: Art and Class in the American Home.* Chicago: University of Chicago Press.

Hanson, Karen. 1986. *The Self Imagined: Philosophical Reflections on the Social Character of the Psyche.* New York: Routledge and Kegan Paul.

Hays, Sharon. 1994. "Structure and Agency and the Sticky Problem of Culture." *Sociological Theory* 12: 57–72.

Hemmer, Helmut. 1990. *Domestication: The Decline of Environmental Appreciation.* Trans. Neil Beckhaus. Cambridge: Cambridge University Press.

Hewitt, John P. 2000. *Self and Society: A Symbolic Interactionist Social Psychology,* 8th ed. Needham Heights, Mass.: Allyn and Bacon.

Hewitt, John P., and Randall G. Stokes. 1975. "Disclaimers." *American Sociological Review* 40: 1–11.

Hochschild, Arlie Russell. 1975. "The Sociology of Feeling and Emotion: Selected Possibilities." Pp. 280–307 in *Another Voice: Feminist Perspectives on Social Life and Social Science,* ed. Marcia Millman and Rosabeth Moss Kanter. Garden City, N.Y.: Anchor Books.

———. 1983. *The Managed Heart: Commercialization of Human Feeling.* Berkeley: University of California Press.

Holstein, James A., and Jaber F. Gubrium. 2000. *The Self We Live By: Narrative Identity in a Postmodern World.* Oxford: Oxford University Press.

Ingold, Tim. 1994. "From Trust to Domination: An Alternative History of Human–Animal Relations." Pp. 1–22 in *Animals and Human Society,* ed. Aubrey Manning and James Serpell. London: Routledge.

Irvine, Leslie. 1997. "Reconsidering the American Emotional Culture: Codependency and Emotion Management." *Innovation: The European Journal of Social Sciences* 10: 345–59.

———. 1999. *Codependent Forevermore: The Invention of Self in a Twelve Step Group.* Chicago: University of Chicago Press.

———. 2000. Even Better than the Real Thing: Narratives of the Self in Codependency." *Qualitative Sociology* 23: 9–28.

———. 2001. "The Power of Play." *Anthrozoös* 14(3):151–60.

———. 2002. "Animal Problems/People Skills: Emotional and Interactional Strategies in Humane Education." *Society & Animals* 10: 63–91.

James, William. 1950 (1890). *The Principles of Psychology.* New York: Dover.

———. 1961 (1892). *Psychology: The Briefer Course.* New York: Harper Torchbooks.

Jamieson, Dale, and Marc Bekoff. 1993. "On Aims and Methods of Cognitive Ethology." *Philosophy of Science Association* 2: 110–24.

Jasper, James M., and Dorothy Nelkin. 1992. *The Animal Rights Crusade: The Growth of a Moral Protest.* New York: Free Press.

Jones, Gareth Stedman. 1971. *Outcast London: A Study in the Relationship between Classes in Victorian Society.* London: Oxford.

Karsh, Eileen B., and Dennis C. Turner. 1988. "The Human–Cat Relationship." Pp. 159–77 in *The Domestic Cat: The Biology of Its Behavior,* ed. D. Turner and P. Bateson. Cambridge: Cambridge University Press.

Katcher, Aaron. 1981. "Interactions between People and Their Pets: Form and Function." Pp. 41–67 in *Interrelations between People and Pets*, ed. Bruce Fogle. Springfield, Ill.: Charles C. Thomas.

Kellert, Stephen R. 1993. "The Biological Basis for Human Values of Nature." Pp. 42–69 in *The Biophilia Hypothesis*, ed. S. Kellert and E. O. Wilson. Washington, D.C.: Island Press/Shearwater Books.

———. 1994. "Attitudes, Knowledge and Behaviour toward Wildlife among the Industrial Superpowers: The United States, Japan and Germany." Pp. 166–87 in *Animals and Human Society: Changing Perspectives*, ed. Aubrey Manning and James Serpell. London: Routledge.

Kete, Kathleen. 1994. *The Beast in the Boudoir: Petkeeping in Nineteenth-Century Paris*. Berkeley: University of California Press.

Kheel, Marti. 1995. "License to Kill: An Ecofeminist Critique of Hunters' Discourse." Pp. 85–125 in *Animals and Women: Feminist Theoretical Explorations*, ed. Carol J. Adams and Josephine Donovan. Durham, N.C.: Duke University Press.

Kidd, Aline H., and Robert M. Kidd. 1980. "Personality Characteristics and Preferences in Pet Ownership." *Psychological Reports* 46: 939–49.

———. 1987. "Seeking a Theory of the Human/Companion Animal Bond." *Anthrozoös* 1: 140–57.

Kruse, Corwin R. 1999. "Gender, Views of Nature, and Support for Animal Rights." *Society & Animals* 7: 179–98.

Lawrence, Elizabeth A. 1986. "Neoteny in American Perceptions of Animals." *Journal of Psychoanalytic Anthropology* 9: 41–54.

———. 1995. "Cultural Perceptions of Differences between People and Animals: A Key to Understanding Human–Animal Relationships." *Journal of American Culture* 18: 75–82.

Lerman, Rhoda. 1996. *In the Company of Newfs*. New York: Henry Holt.

Lerner, Jennifer E., and Linda Kalof. 1999. "The Animal Text: Message and Meaning in Television Advertisements." *Sociological Quarterly* 40: 565–86.

Leyhausen, P. 1979. *Cat Behavior: The Predatory and Social Behavior of Domestic and Wild Cats*. Trans. B. A. Tonkin. New York: Garland.

Linden, Eugene. 1999. *The Parrot's Lament, and Other True Tales of Animal Intrigue, Intelligence, and Ingenuity*. New York: Penguin Putnam.

Linzey, Andrew. 1998. "Christianity." Pp. 286–88 in *Encyclopedia of Animal Rights and Animal Welfare*, ed. Marc Bekoff. Westport, Conn.: Greenwood.

Maehle, Andreas-Holger. 1994. "Cruelty and Kindness to the 'Brute Creation': Stability and Change in the Ethics of the Man–Animal Relationship, 1600–1850." Pp. 81–105 in *Animals and Human Society: Changing Perspectives*, ed. Aubrey Manning and James Serpell. London: Routledge.

Maggitti, Paul. 1996. "Cat Litter: The Inside Scoop." *Pet Business* (July).

Málek, Jaromír. 1993. *The Cat in Ancient Egypt*. London: British Museum Press.

Martin, Paul, and Patrick Bateson. 1988. "Behavioural Development in the Cat." Pp. 9–22 in *The Domestic Cat: The Biology of Its Behaviour,* ed. Dennis C. Turner and Patrick Bateson. Cambridge: Cambridge University Press.

Masson, Jeffrey Moussiaeff. 1997. *Dogs Never Lie about Love: Reflections on the Emotional World of Dogs.* New York: Three Rivers Press.

Masson, Jeffrey Moussaieff, and Susan McCarthy. 1995. *When Elephants Weep: The Emotional Lives of Animals.* New York: Delta.

McDonogh, Kathleen. 1999. *Reigning Cats and Dogs.* New York: St. Martin's Press.

Mead, George Herbert. 1962 (1934). *Mind, Self and Society.* Chicago: University of Chicago Press.

Melson, Gail F. 2001. *Why the Wild Things Are: Animals in the Lives of Children.* Cambridge, Mass.: Harvard University Press.

Menache, Sophia. 1997. "Dogs: God's Worst Enemies?" *Society & Animals* 5: 23–44.

———. 2000. "Hunting and Attachment to Dogs in the Pre-Modern Period." Pp. 42–60 in *Companion Animals and Us: Exploring the Relationships between People and Pets,* ed. Anthony L. Podberscek, Elizabeth S. Paul, and James A. Serpell. Cambridge: Cambridge University Press.

Mertens, Claudia. 1991. "Human–Cat Interactions in the Home Setting." *Anthrozoös* 4: 224–33.

Messent, Peter. 1983. "Social Facilitation of Contact with Other People by Pet Dogs." Pp. 37–46 in *New Perspectives on Our Lives with Companion Animals,* ed. Aaron Katcher and Alan Beck. Philadelphia: University of Pennsylvania Press.

Messent, Peter R., and James A. Serpell. 1981. "An Historical and Biological View of the Pet–Owner Bond." Pp. 5–22 in *Interrelations between People and Pets,* ed. Bruce Fogel. Springfield, Ill.: Charles C. Thomas.

Midgley, Mary. 1983. *Animals and Why They Matter.* Athens: University of Georgia Press.

Milekic, Slavoljub. 1998. "Disneyfication." Pp. 133–34 in *Encyclopedia of Animal Rights and Animal Welfare,* ed. Marc Bekoff. Westport, Conn.: Greenwood.

Mills, C. Wright. 1940. "Situated Actions and Vocabularies of Motive." *American Sociological Review* 5: 904–13.

Mitchell, Robert W., Nicholas S. Thompson, and H. Lyn Miles, eds. 1997. *Anthropomorphism, Anecdotes, and Animals.* Albany: State University of New York Press.

Morris, Paul, Margaret Fidler, and Alan Costall. 2000. "Beyond Anecdotes: An Empirical Study of 'Anthropomorphism.'" *Society & Animals* 8: 151–65.

Myers, Gene. 1998. *Children and Animals: Social Development and Our Connections to Other Species.* Boulder, Colo.: Westview Press.

Neisser, Ulric. 1994. "Self-Narratives: True and False." Pp. 1–18 in *The Remembering Self: Construction and Accuracy in the Self-Narrative*, ed. Ulric Neisser and Robyn Fivush. London: Cambridge University Press.

Nibert, David A. 1994. "Animal Rights and Human Social Issues." *Society & Animals* 2: 115–24.

———. 2002. *Animal Rights/Human Rights: Entanglements of Oppression and Liberation*. Lanham, Md.: Rowman and Littlefield.

Niven, Charles D. 1967. *History of the Humane Movement*. London: Johnson.

Noske, Barbara. 1997. *Beyond Boundaries: Humans and Animals*. Montreal: Black Rose Books.

Owens, Paul, and Norma Eckroate. 1999. *The Dog Whisperer: A Compassionate, Nonviolent Approach to Dog Training*. Holbrook, Mass.: Adams Media.

Parsons, Michael J. 1987. *How We Understand Art: A Cognitive-Developmental Account of Aesthetic Response*. New York: Cambridge University Press.

Patterson, Francine, and Eugene Linden. 1981. *The Education of Koko*. New York: Holt, Rinehart, and Winston.

Pepperberg, Irene. 1991. "A Communicative Approach to Animal Cognition: A Study of Conceptual Abilities of an African Grey Parrot." Pp. 153–86 in *Cognitive Ethology: The Minds of Other Animals*, ed. Carolyn A. Ristau. Hillsdale, N.J.: Lawrence Erlbaum.

Perin, Constance. 1981. "Dogs as Symbols in Human Development." Pp. 68–88 in *Interrelations between People and Pets*, ed. Bruce Fogle. Springfield, Ill.: Charles C. Thomas.

Pfungst, Otto. 1911. *Clever Hans: A Contribution to Experimental Animal and Human Psychology*. New York: Henry Holt.

Phillips, Mary T. 1994. "Proper Names and the Social Construction of Biography: The Negative Case of Laboratory Animals." *Qualitative Sociology* 17: 119–42.

Plous, Scott. 1993. "Psychological Mechanisms in the Human Use of Animals." *Journal of Social Issues* 49: 11–52.

Plummer, Ken. 1983. *Documents of Life: An Introduction to the Problems and Literature of a Humanistic Method*. London: Allen and Unwin.

Podberscek, Anthony, and Samuel D. Gosling. 2000. "Personality Research on Pets and Their Owners: Conceptual Issues and Review." Pp. 143–67 in *Companion Animals and Us: Exploring the Relationships between People and Pets*, ed. Anthony L. Podberscek, Elizabeth S. Paul, and James A. Serpell. Cambridge: Cambridge University Press.

Podberscek, Anthony L. Elizabeth S. Paul, and James A. Serpell. 2000. *Companion Animals and Us: Exploring the Relationships between People and Pets*. Cambridge: Cambridge University Press.

Pollner, Melvin, and Lynn McDonald-Wikler. 1985. "The Social Construction of Unreality: A Case Study of a Family's Attribution of Competence to a Severely Retarded Child." *Family Process* 24: 241–54.

Poresky, Robert H., Charles Hendrix, Jacob E. Moser, and Marvin L. Samuelson. 1988. "Young Children's Companion Animal Bonding and Adults' Pet Attitudes: A Retrospective Study." *Psychological Reports* 62: 419–25.

Posage, J. Michelle, Paul C. Bartlett, and Daniel K. Thompson. 1998. "Determining Factors for Successful Adoption of Dogs from an Animal Shelter." *Journal of the American Veterinary Medical Association* 213: 478–82.

Pryor, Karen. 1986. *Don't Shoot the Dog: The New Art of Teaching and Training.* New York: Bantam.

Regan, Tom. 1983. *The Case for Animal Rights.* Berkeley: University of California Press.

Ristau, Carolyn R., ed. 1991. *Cognitive Ethology: The Minds of Other Animals.* Hillsdale, N.J.: Lawrence Erlbaum.

Ritvo, Harriet. 1987. *The Animal Estate: The English and Other Creatures in the Victorian Age.* Cambridge, Mass.: Harvard University Press.

———. 1988. "The Emergence of Modern Pet-Keeping." Pp. 13–31 in *Animals and People Sharing the World,* ed. Andrew N. Rowan. Hanover, N.H.: University Press of New England.

Roberts, William A., and Dwight S. Mazmanian. 1988. "Concept Learning at Different Levels of Abstraction by Pigeons, Monkeys and People." *Journal of Experimental Psychology: Animal Behavior Processes* 14: 247–60.

Robins, Douglas M., Clinton R. Sanders, and Spencer E. Cahill. 1991. "Dogs and Their People: Pet-Facilitated Interaction in a Public Setting." *Journal of Contemporary Ethnography* 20: 3–25.

Rollin, Bernard E. 1992. *Animal Rights and Human Morality,* 2nd ed. Buffalo, N.Y.: Prometheus Books.

Rollin, Bernard E., and Michael D. H. Rollin. 2001. "Dogmatisms and Catechisms: Ethics and Companion Animals." *Anthrozoös* 14: 4–11.

Rubenstein, David. 2001. *Culture, Structure, and Agency: Toward a Truly Multidimensional Sociology.* Thousand Oaks, Calif.: Sage.

Saint-Exupéry, Antoine de. 1971 (1943). *The Little Prince.* Trans. Richard Howard. San Diego: Harvest/Harcourt.

Sanders, Clinton R. 1990. "Excusing Tactics: Social Responses to the Public Misbehavior of Companion Animals." *Anthrozoös* 4: 82–90.

———. 1991. "The Animal 'Other': Self-Definition, Social Identity, and Companion Animals." Pp. 662–68 in *Advances in Consumer Research,* vol. 17, ed. Marvin Goldberg et al. Provo, Utah: Association for Consumer Research.

———. 1993. "Understanding Dogs: Caretakers' Attributions of Mindedness in Canine–Human Relationships." *Journal of Contemporary Ethnography* 22: 205–26.

———. 1994a. "Annoying Owners: Routine Interactions with Problematic Clients in a General Veterinary Practice." *Qualitative Sociology* 17: 159–70.

———. 1994b. "Biting the Hand That Heals You: Encounters with Problematic Patients in a General Veterinary Practice." *Society & Animals* 2: 47–66.

———. 1999. *Understanding Dogs: Living and Working with Canine Companions.* Philadelphia: Temple University Press.

———. 2000. "The Impact of Guide Dogs on the Identity of People with Visual Impairments." *Anthrozoös* 13: 131–39.

Sanders, Clinton R., and Arnold Arluke. 1993. "If Lions Could Speak: Investigating the Animal–Human Relationship and the Perspectives of Nonhuman Others." *Sociological Quarterly* 34: 377–90.

Sauer, Carl Ortwin. 1952. *Agricultural Origins and Dispersals.* Cambridge, Mass.: MIT Press.

Schoen, Allen M. 2001. *Kindred Spirits.* New York: Broadway Books/Random House.

Schwabe, Calvin. 1994. "Animals in the Ancient World." Pp. 36–58 in *Animals and Human Society: Changing Perspectives,* ed. Aubrey Manning and James Serpell. London: Routledge.

Scully, Matthew. 2002. *Dominion: The Power of Man, the Suffering of Animals, and the Call to Mercy.* New York: St. Martin's Press.

Serpell, James. 1981."Childhood Pets and Their Influence on Adults' Attitudes." *Psychological Reports* 49:651–654.

———. 1986. *In the Company of Animals.* Oxford: Basil Blackwell.

———. 1988a. "Pet-Keeping in Non-Western Societies: Some Popular Misconceptions." Pp. 34–52 in *Animals and People Sharing the World,* ed. Andrew N. Rowan. Hanover, N.H.: University Press of New England.

———. 1988b. "The Domestication and History of the Cat." Pp. 151–58 in *The Domestic Cat: The Biology of Its Behaviour,* ed. Dennis C. Turner and Patrick Bateson. Cambridge: Cambridge University Press.

———. 1995. "From Paragon to Pariah: Some Reflections of Human Attitudes to Dogs." Pp. 245–56 in *The Domestic Dog: Its Evolution, Behaviour and Interactions with People,* ed. James Serpell. Cambridge: Cambridge University Press.

Shapiro, Kenneth J. 1990. "Understanding Dogs through Kinesthetic Empathy, Social Construction, and History." *Anthrozoös* 3: 184–95.

———. 1997. "A Phenomenological Approach to the Study of Nonhuman Animals." Pp. 277–95 in *Anthropomorphism, Anecdotes, and Animals,* ed. R. Mitchell, N. Thompson and H. Miles. Albany: State University of New York Press.

Sheldrake, Rupert. 1999. *Dogs That Know When Their Owners Are Coming Home.* New York: Crown.

Sheldrake, Rupert, and Marc Bekoff. 2000. "In Conversation." *The Bark: The Modern Dog Culture Magazine,* no. 10 (Winter), 48–50.

Shepard, Paul. 1978. *Thinking Animals: Animals and the Development of Human Intelligence*. New York: Viking Press.

———. 1996. *The Others: How Animals Made Us Human*. Washington, D.C.: Island Press.

Shue, Henry. 1996. *Basic Rights*, 2nd ed. Princeton, N.J.: Princeton University Press.

Siegel, Judith M. 1993. "Companion Animals: In Sickness and in Health." *Journal of Social Issues* 49: 157–67.

Siegal, Mordecai, ed. 1989. *The Cornell Book of Cats*. New York: Villard.

Singer, Peter. 1990. *Animal Liberation*, rev. ed. New York: New York Review of Books.

Sommers, Shula. 1984. "Adults Evaluating Their Emotions: A Cross-Cultural Perspective." Pp. 319–38 in *Emotion and Adult Development*, ed. Carol Zander Malatesta and Carroll E. Izard. Beverly Hills, Calif.: Sage.

Sorabji, Richard. 1993. *Animal Minds and Human Morals: The Origins of the Western Debate*. Ithaca, N.Y.: Cornell University Press.

Sperling, Susan. 1988. *Animal Liberators: Research and Morality*. Berkeley: University of California Press.

Spiegel, Ian Brett. 2000. "A Pilot Study to Evaluate an Elementary School-Based Dog Bite Prevention Program." *Anthrozoös* 13: 164–73.

Stearns, Peter N. 1989a. "Suppressing Unpleasant Emotions: The Development of a Twentieth-Century American Style." Pp. 230–61 in *Social History and Issues in Human Consciousness: Some Interdisciplinary Connections*, ed. Andrew E. Burns and Peter N. Stearns. New York: New York University Press.

———. 1989b. *Jealousy: The Evolution of an Emotion in American History*. New York: New York University Press.

———. 1994. *American Cool*. New York: New York University Press.

Stern, Daniel N. 1985. *The Interpersonal World of the Infant: A View from Psychoanalysis and Developmental Psychology*. New York: Basic Books.

Stevens, Montague. 1990 (1943). *Meet Mr. Grizzly*. Silver City, N.M.: High Lonesome Books.

Strauss, Anselm, ed. 1964. *George Herbert Mead on Social Psychology*. Chicago: University of Chicago Press.

Swabe, Joanna. 2000. "Veterinary Dilemmas: Ambiguity and Ambivalence in Human–Animal Interaction." Pp. 292–312 in *Companion Animals and Us: Exploring the Relationships between People and Pets*, ed. Anthony L. Podberscek, Elizabeth S. Paul, and James A. Serpell. Cambridge: Cambridge University Press.

Tabor, Roger. 1983. *The Wild Life of the Domestic Cat*. London: Arrow Books.

Tester, Keith. 1992. *Animals and Society: The Humanity of Animal Rights*. London: Routledge.

Thomas, Elizabeth Marshall. 1993. *The Hidden Life of Dogs.* Boston and New York: Houghton Mifflin.

———. 1994. *The Tribe of Tiger: Cats and Their Culture.* New York: Simon and Schuster.

———. 2000. *The Social Lives of Dogs: The Grace of Canine Company.* New York: Simon and Schuster.

Thomas, Keith. 1983. *Man and the Natural World: Changing Attitudes in England 1500–1800.* London: Allen Lane.

Tuan, Yi-Fu. 1984. *Dominance and Affection: The Making of Pets.* New Haven, Conn.: Yale University Press.

Turner, James. 1980. *Reckoning with the Beast: Animals, Pain, and Humanity in the Victorian Mind.* Baltimore: Johns Hopkins University Press.

Unti, Bernard. 1998. "Caroline White." P. 362 in *Encyclopedia of Animal Rights and Animal Welfare,* ed. Marc Bekoff. Westport, Conn.: Greenwood.

Voltaire. (1962). *Philosophical Dictionary,* translated with an introduction and glossary by Peter Gay, vol. 1. New York: Basic Books.

Weber, Max. 1954. *The Protestant Ethic and the Spirit of Capitalism.* New York: Free Press

———. 1968 (1922). *Economy and Society: An Outline of Interpretive Sociology.* New York: Bedminster Press.

Wells, Deborah L., and Peter G. Hepper. 1999. "Male and Female Dogs Respond Differently to Men and Women." *Applied Animal Behaviour Science* 60: 83–88.

———. 2001. "The Behavior of Visitors towards Dogs Housed in an Animal Rescue Shelter." *Anthrozoös* 14: 12–18.

West, Candace. 1999. "Not Even a Day in the Life." Pp. 3–12 in *Qualitative Sociology as Everyday Life,* ed. Barry Glassner and Rosanna Hertz. Thousand Oaks, Calif.: Sage.

Wilson, Cindy C. and Dennis C. Turner. 1998. *Companion Animals in Human Health.* Thousand Oaks, Calif.: Sage.

Wilson, Edward O. 1993. "Biophilia and the Conservation Ethic." Pp. 31–41 in *The Biophilia Hypothesis,* ed. S. Kellert and E. O. Wilson. Washington, D.C.: Island Press/Shearwater Books.

Winnicott, D. W. 1958. *Collected Papers.* London: Tavistock.

Wise, Steven M. 2000. *Rattling the Cage: Toward Legal Rights for Animals.* Cambridge, Mass.: Perseus.

Wolch, Jennifer. 1998. "Zoöpolis." Pp. 119–38 in *Animal Geographies: Place, Politics, and Identity in the Nature–Culture Borderlands,* ed. Jennifer Wolch and Jody Emel. London: Verso.

Wouters, Cas. 1991. "On Status Competition and Emotion Management." *Journal of Social History* 24: 699–717.

Index

About the Authors

Leslie Irvine is Assistant Professor of Sociology at the University of Colorado, Boulder, and the author of *Codependent Forevermore: The Invention of Self in a Twelve Step Group*.

Marc Bekoff (home page: http://literati.net/Bekoff) teaches biology at the University of Colorado, Boulder. He is the author or editor of many books, including the *Encyclopedia of Animal Rights and Animal Welfare*, *Strolling with Our Kin*, *The Smile of a Dolphin: Remarkable Accounts of Animal Emotions*, *Minding Animals: Awareness, Emotions, and Heart*, and *The Ten Trusts* (with Jane Goodall). He and Dr. Goodall recently co-founded Ethologists for the Ethical Treatment of Animals/Citizens for Responsible Animal Behavior Studies (http://www.ethologicalethics.org).